Lessons in Scientific Computing

Numerical Mathematics, Computer Technology, and Scientific Discovery

Lessons in Scientific Computing

Numerical Mathematics, Computer Technology, and Scientific Discovery

By

Norbert Schörghofer

CRC Press
Taylor & Francis Group
Boca Raton London New York

CRC Press is an imprint of the
Taylor & Francis Group, an **informa** business

CRC Press
Taylor & Francis Group
6000 Broken Sound Parkway NW, Suite 300
Boca Raton, FL 33487-2742

© 2019 by Taylor & Francis Group, LLC
CRC Press is an imprint of Taylor & Francis Group, an Informa business

No claim to original U.S. Government works

Printed on acid-free paper
Version Date: 20180411

International Standard Book Number-13: 978-1-138-07063-9 (Hardback)

**Visit the Taylor & Francis Web site at
http://www.taylorandfrancis.com**

**and the CRC Press Web site at
http://www.crcpress.com**

Contents

Preface

Fundamental scientific discoveries have been made, and continue to be made, with the help of computational methods. For example, commonalities in the behavior of chaotic systems, most prominently Feigenbaum universality, have been discovered by means of numerical calculations. This required only simple and small computations. An example at the opposite extreme, using large and complex computations, is the prediction of the mass of the proton from fundamental theories of physics; only with numerical calculations is it possible to evaluate the theoretical expressions with sufficient accuracy to allow comparison with experimental measurement. Such examples illustrate the crucial role of numerical calculations in basic science.

Before the dawn of electronic computers, Milutin Milankovitch spent one hundred full days, from morning until night, with paper and fountain pen calculations on one of his investigations into the astronomical theory of Earth's Ice Ages. Today, the same computation could be performed within seconds with electronic computers. Even small numerical calculations can solve problems not easily accessible with mathematical theory.

While today's students and researchers spend a great portion of their time on computational modeling, comparatively little educational material is available about the computational branch of scientific inquiry. Moreover, the rapid advances in computer technology have changed the way we do scientific computing. As a consequence of such developments, part of the conventional wisdom, and many existing textbooks, have become outdated.

This book is a modernized, broad, and compact introduction into scientific computing. It combines the various components of the field (numerical analysis, discrete numerical mathematics, computer science, and computational hardware), subjects that are most often taught separately, into one book. The book takes a broad and interdisciplinary approach. Numerical methods, computer technology, and their interconnections are treated with the goal to facilitate scientific research. Aspects from several subject areas are often compounded within a single chapter. Combining a vast universe of relevant knowledge into one book, short enough so it can be digested over the course of a semester at a university, required severe prioritizations. Often when working on the manuscript, it grew shorter, because less relevant material was discarded. It is written with an eye on breadth, usefulness, and longevity.

The book covers fundamental concepts and modern practice. It is aimed at graduate courses and advanced undergraduate courses in scientific computing from multiple disciplines, including physics, chemistry, astronomy, and for all those who have passed through standard core quantitative classes in their undergraduate years. Sections with a star symbol (*) contain specialized or advanced material. In the text, a triple >>> indicates a Python prompt, and a single > a Unix or other type of prompt. *Italics* indicate emphasis or paragraph headings. `Truetype` is used for program commands, program variables, and binary numbers. For better readability, references and footnotes within the text are omitted. Brainteasers at the end of several of the chapters should stimulate discussion.

I am indebted to my former teachers, colleagues, and to many textbook authors who have, knowingly or unknowingly, contributed to the substance of this book. They include Professor Leo Kadanoff, who taught me why we compute, Professor James Sethian, on whose lectures chapter 15 is based on, and the authors of "Numerical Recipes," whose monumental and legendary book impacted my student years. Several friends and colleagues have provided valuable feedback on draft versions of this material: Mike Gowanlock, Geoff Mathews, and Tong Zhou, and I am indebted to them.

NORBERT SCHÖRGHOFER

Honolulu, Hawaii
March 2018

To the Instructor

This book offers a modernized, broad, and compact introduction into scientific computing. It is appropriate as a main textbook or as supplementary reading in upper-level college and in graduate courses on scientific computing or discipline-specific courses on computational physics, astrophysics, chemistry, engineering, and other areas where students are prepared through core quantitative classes. Prerequisites are basic calculus, linear algebra, and introductory physics. Occasionally, a Taylor expansion, an error propagation, a Fourier transform, a matrix determinant, or a simple differential equation appears. It is recommended that an undergraduate course omit the last two chapters and some optional sections, whereas a graduate course would cover all chapters and at a faster pace.

Prior knowledge of numerical analysis and a programming language is optional, although students will have to pick up programming skills during the course. In today's time and age, it is reasonable to expect that every student of science or engineering is, or will be, familiar with at least one programming language. It is easy to learn programming by example, and simple examples of code are shown from several languages, but the student needs to be able to write code in only one language. Some books use pseudocode to display programs, but fragments of C or Python code are as easy to understand.

Supplementary material is available at `https://github.com/nschorgh/CompSciBook/`. This online material is not necessary for the instructor or the student, but may be useful nevertheless.

The lectures can involve interactive exercises with a computer's display projected in class. These interactive exercises are incorporated in the text, and the corresponding files are included in the online repository.

The book offers a grand tour through the world of scientific computing. By the end, the student has been exposed to a wide range of methods, fundamental concepts, and practical material. Even some graduate-level science is introduced in optional sections, such as almost integrable systems, diagrammatic perturbation expansions, and electronic density functionals.

Among the homework problems, three fit with the presentation so well that it is recommended to assign them in every version of the course. The "invisible roots" of Exercise 2.1 could be demonstrated in class, but the surprise may be bigger when the student discovers them. For Exercise 3.1, every member of the class, using different computing platforms and programming languages, should be able to exactly reproduce a highly round-off sensitive result. In Exercise 5.1, the students themselves can answer the qualitative question whether a kicked rotator can accumulate energy indefinitely. By fitting the asymptotic dependence of energy with time, some will likely use polynomials of high degree, a pitfall that will be addressed in the subsequent chapter.

If instructors would send me the result of the survey of computing background, posted as an exercise at the end of the first chapter, I could build a more representative data set.

Behind the Scenes: Design of a Modern and Integrated Course

The subject of scientific computing can be viewed to stand on three pillars: 1) numerical mathematics, 2) programming and computer hardware, and 3) computational problem solving for the purpose of scientific research. This book seeks to integrate all of these aspects.

An integrated approach has two fundamental advantages. First it allows the subject to be taught as a one-stop shop. Few students find the time to take individual courses in numerical mathematics, programming, and scientific modeling. Second, an integrated approach allows each topic to be placed in a broader context. For example, the operation count of numerical mathematics can be appreciated to a greater depth when viewed on the basis of the speed of the physical components of the hardware.

For any course, the selection of topics, what to teach versus what to leave out, is a crucial responsibility of the instructor or book author, and the planning of the content was a major component of the work that led to this book. The design of the content began with a list of ranked priorities, and this book should be judged in part by its topic selection and by what it does *not* contain.

To make efficient use of the reader's time, each chapter often describes methods and simultaneously covers a subject area. For example, in chapter 1, binary floating-point numbers are introduced during a description of chaotic maps. In chapter 2, methods of root finding lead to the concept of problems without guaranteed solution. The concept of the operation count is introduced with application to numerical linear algebra, and so on.

The advances in electronic computers and the broad availability of implementations for all common numerical problems permeates into how scientific modeling is done in daily life. Whereas, when fitting an interpolating polynomial, one was once concerned with how to obtain these coefficients, now the relevant concern is more along the lines of whether the approximating function is well-behaved. Whereas, one used to be most concerned with computational efficiency, it is now the pace and reliability of the coding effort that is the most common concern. Whereas, every classical course on numerical analysis would cover numerical methods for ordinary differential equations (ODEs), numerically integrating a few ODEs is rarely a challenge today; therefore these numerical methods are described only briefly here, and priority is given to numerical methods for partial differential equations.

To arrive at a selection of science applications, I compiled a list of examples where computations led to a major breakthrough, and then evaluated which of these examples could be treated in an introductory textbook in a sufficiently simplified form. The standard map of Hamiltonian chaos offers an ideal combination of simple programming yet complex and instructive results. Among potential examples of algorithmically intractable problems is multiple genome alignment, because of its relevance in bioinformatics. The diagrammatic Mayer cluster expansion, an optional section on symbolic computing, is not chosen because the Mayer cluster expansion is so important per-se, but because the concept of high-order diagrammatic perturbation expansions is important in quantum field theory.

A significant habitual problem in scientific computing is that the novelties in the world of computing act as a distraction from the scientific goal. The book intentionally begins and ends with examples where computations truly made a crucial difference: the Feigenbaum universality of period doubling in chapter 1 and an outline of the Density Functional Method in chapter 16. Several science-driven examples are embedded throughout the text, and part of their purpose is to keep the focus on computational problem-solving. Two applications that appear repeatedly are chaotic systems and gravitational N-body interaction. They are revisited not because they would be more important than other applications, but because each could be used for many purposes. For a while I was torn whether the three-body problem should be included in chapter 5, but when the first interstellar visitor happened to be discovered while working on this chapter, I knew it belonged.

As in all areas, the content of textbooks lags behind modern practice. Much of classical numerical mathematics focuses nearly exclusively on computational efficiency (rather than on coding or data transfer bandwidth), and many programming curricula are written for single-core computations. This course modernizes the curriculum by emphasizing coding practices and parallel computing.

One challenge, for any book on this subject, is to keep up to date with the rapidly changing technology, both hardware and software. The following are components that I would not have been prudent enough to include a decade earlier: 1) Python, which has captured such a large share of scientific computing, is featured prominently in the book, compared to other languages; 2) a dedicated chapter on data; and 3) pervasive treatment of concurrent computations, appropriate for modern (intrinsically parallel) CPUs. And to prepare students for future technology changes, the approach was to provide a broad perspective of programming languages and computing hardware and to explain how technology impacts computing.

## CHAPTER 1

Analytical & Numerical Solutions

Many equations describing the behavior of physical systems cannot be solved analytically. In fact, "most" can not. Solutions can still be obtained numerically, and that opens up tremendous possibilities. Numerical methods can be useful because they provide quantitative answers that would otherwise elude us. But numerical results not only provide quantitative answers, they can also provide new insight. This chapter presents one such example of how numerical calculations are used to gain insight.

1.1 NUMERICAL EXPLORATION

A a short computer program suffices for a simple demonstration. If we repeatedly evaluate $x_{n+1} = x_n(1 - x_n)$, starting with an arbitrary value $0 < x < 1$, the number will decrease and slowly approach zero. For example,

$$x = 0.800, \, 0.160, \, 0.134, \, 0.116, \, 0.103, \ldots$$

where only the first three significant digits are shown. The sequence decreases because $x(1 - x) < x$ for any $x > 0$. If we try instead $x_{n+1} = 3.1x_n(1 - x_n)$, the iteration is no longer driven toward a constant. For example,

$$x = 0.800, \, 0.496, \, 0.775, \, 0.541, \, 0.770, \ldots$$

A few iterations later

$$\ldots, \, 0.765, \, 0.558, \, 0.765, \, 0.558, \, 0.765, \, 0.558, \ldots$$

The iteration settles into a periodic behavior; it alternates between two values. A yet higher prefactor, $x_{n+1} = 4x_n(1 - x_n)$, produces

$$x = 0.800, \, 0640, \, 0.922, \, 0.289, \, 0.822, \ldots$$

One thousand iterations later

$\ldots, 0.456, 0.992, 0.031, 0.122, 0.428, 0.979, 0.082, 0.301, 0.841, \ldots$

which has no recognizable period (Figure 1.1). We are faced with a dilemma on how to interpret this numerical result and whether to trust it. Perhaps roundoff alters the behavior? Is the period unrecognizably long? Do the numbers eventually settle into a regular behavior, but the transient is long? Or does this iteration not approach anything at all?

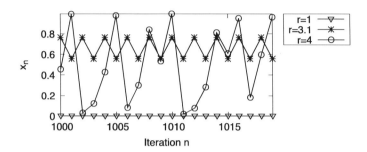

FIGURE 1.1   Behavior of iteration $x_{n+1} = rx_n(1 - x_n)$ with $x_0 = 0.8$ after many iterations for three values of the parameter r.

Can it be true that the iteration does not settle to a constant or into a periodic pattern, or is this an artifact of numerical inaccuracies? Consider the simple iteration

$$y_{n+1} = 1 - |2y_n - 1|$$

known as "tent map." For $y_n \leq 1/2$ the value is doubled, $y_{n+1} = 2y_n$, and for $y_n \geq 1/2$ it is subtracted from 1 and then doubled, $y_{n+1} = 2(1 - y_n)$. The equation can equivalently be written as

$$y_{n+1} = \begin{cases} 2y_n & \text{for} \quad 0 \leq y_n \leq 1/2 \\ 2(1 - y_n) & \text{for} \quad 1/2 \leq y_n \leq 1 \end{cases}$$

Figure 1.2 illustrates this function, along with the iteration $x_{n+1} = 4x_n(1 - x_n)$, used above. The latter is known as "logistic map."

The behavior of the tent map is particularly easy to understand when y_n is represented in the binary number system. As for integers, floating-point numbers can be cast in binary representation. For example, the integer binary number 10 is $1 \times 2^1 + 0 \times 2^0 = 2$. The binary floating-point number 0.1 is $1 \times 2^{-1} = 0.5$ in decimal system, and binary 0.101 is $1 \times 2^{-1} + 0 \times 2^{-2} + 1 \times 2^{-3} = 0.625$.

For binary numbers, multiplication by two corresponds to a shift by one digit, just as multiplication by 10 shifts any decimal number by one digit. The tent map changes 0.011 to 0.110. When a binary sequence is subtracted from 1, zeros and ones are simply interchanged, as can be seen from the following argument. Given the identity $\sum_{j=1}^{\infty} 2^{-j} = 1$, 1 is equivalent to 0.11111...

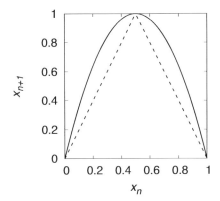

FIGURE 1.2 The functions $4x(1-x)$ and $1 - |2x - 1|$.

with infinitely many 1s. From that it is apparent that upon subtraction from 1, zeros and ones are swapped. For example, `1-0.01` is

```
  0.1111111...
- 0.0100000
  0.1011111...   (digits of 0.0100000 swapped)
```

where the subtraction was done as if this was elementary school. As a last step, the identity `0.0011111...` = `0.01` could be used again. So in total, `1-0.01` = `0.11`, which is true, because $1 - 1/4 = 1/2 + 1/4$.

Armed with this knowledge, it is easy to see how the tent map transforms numbers. The iteration goes from `0.011001...` (which is $< 1/2$) to `0.11001...` ($> 1/2$) to `0.0110....` After each iteration, numbers advance by one digit, because of multiplication by 2 for either half of the tent map. After many iterations the digits from far behind dominate the result. Hence, the leading digits take on values unrelated to the leading digits of earlier iterations, making the behavior of the sequence apparently random.

This demonstrates that a simple iteration can produce effectively random behavior. We conclude that it is plausible that a simple iteration never settles into a periodic behavior. The tent map applied to a binary floating-point number reveals something mathematically profound: deterministic equations can exhibit random behavior, a phenomenon known as "deterministic chaos."

The substitution $x_n = \sin^2(\pi y_n)$ transforms the tent map into the logistic map. This follows from trigonometry:

$$
\begin{aligned}
x_{n+1} &= 4x_n(1 - x_n) \\
\sin^2(\pi y_{n+1}) &= 4\sin^2(\pi y_n)(1 - \sin^2(\pi y_n)) \\
&= 4\sin^2(\pi y_n)\cos^2(\pi y_n) = \sin^2(2\pi y_n)
\end{aligned}
$$

We arrive at the relation $|\sin(\pi y_{n+1})| = |\sin(2\pi y_n)|$. When $0 \leq y_n \leq 1/2$, this equates to $y_{n+1} = 2y_n$. For $1/2 \leq y_n \leq 1$, the sine on the right-hand side is negative, so in this range $\sin(\pi y_{n+1}) = -\sin(2\pi y_n) = \sin(\pi - 2\pi y_n)$, or $y_{n+1} = 1 - 2y_n$. Hence, the logistic map can be transformed back to the tent map, whose chaotic behavior is proven mathematically. This transformation proves the logistic map is indeed chaotic. And it answers how the iteration behaves when treated at infinite accuracy: the iteration does not settle to a constant value nor into any periodic pattern.

It is remarkable that the numerical solution of the logistic map maintains the stochastic character. Whereas it would be reasonable to expect roundoff might erode the stochasticity of a solution, this is not apparent even after a great many iterations. That said, here is a final twist: Numerical simulation of the tent map *is* hampered by roundoff; it eventually settles on zero. This incidentally illustrates the range of possibilities for the interaction between computational results and the human mind: Numerics can be right when we think it is wrong, and it can be wrong when we think it is right.

1.2 A COMPUTATIONAL DISCOVERY: UNIVERSALITY OF PE- RIOD DOUBLING

The asymptotic solutions of the iteration formulae $x_{n+1} = rx_n(1 - x_n)$, with increasing parameter r, are readily visualized by plotting the value of x for many iterations. The initial transient is eliminated by discarding the first thousand or so iterations. If x approaches a constant, then the result consists of only one point; if it becomes periodic, there will be several points, and if it is chaotic (nonperiodic), there will be a range of values. Figure 1.3(a) shows the asymptotic behavior for the range of parameter values $r = 0$ to $r = 4$. As we have seen in the examples above, the asymptotic value for $r = 1$ is zero; $r = 3.1$ settles into a period of two, and for $r = 4$ the behavior is chaotic. With increasing r the period doubles repeatedly, from 1 to 2 to 4 to 8 and on, and then the iteration transitions into chaos. (Figure 1.3(b) shows an enlarged version of a portion of Figure 1.3(a).) The chaotic parameter region is interrupted by windows of periodic behavior; these windows turn out to have odd rather than even periods. Although the iteration is simple, its asymptotic solutions are truly complex.

So far, numerics has enabled us to explore the complex behavior of the logistic map, which includes period doubling and chaotic solutions. A monumental realization comes when this behavior is compared with solutions of other iterative equations. Figure 1.4 shows a similar iteration, the "sine map" $x_{n+1} = \sin(rx_n)$, which also exhibits period doubling and chaos as r increases from 0 to π. Many, many other iterative equations show the same behavior of period doubling and chaos. The generality of this phenomenon, called Feigenbaum universality, was not realized before the numerical computations came along.

(The universality of the period doubling behavior goes beyond the qualitative similarities between Figures 1.3(a) and 1.4; it also has specific quantitative aspects. The spacing between the critical parameter values r_n of the

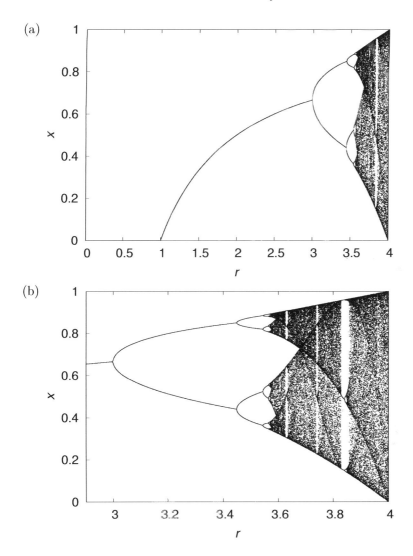

FIGURE 1.3 Asymptotic behavior of the iterative equation $x_{n+1} = rx_n(1-x_n)$ with varying parameter r. The iteration converges to a fixed value for $r \leq 1$, exhibits periodic behavior with periods 2, 4, 8, ..., and becomes chaotic around $r \approx 3.57$. Panel (b) shows an enlarged portion of panel (a).

bifurcations to cycles of period 2^n decreases, and it turns out to decrease like a geometric sequence. The ratio $(r_n - r_{n-1})/(r_{n+1} - r_n)$ approaches a specific constant $\lim_{n\to\infty}(r_n - r_{n-1})/(r_{n+1} - r_n) = 4.6692\ldots$ — the very

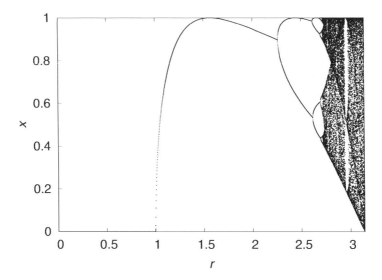

FIGURE 1.4 Asymptotic behavior for the iteration $x_{n+1} = \sin(rx_n)$ with varying parameter r. This iteration also exhibits period doubling and chaotic behavior. Compare with Figure 1.3(a).

same constant for many iterative equations. It is known as the first Feigenbaum constant. The second Feigenbaum constant asymptotically describes the shrinking of the width of the periodic cycles. If d_n is the distance between adjacent x-values of the cycle of period 2^n, then d_n/d_{n+1} converges to $2.5029\ldots$)

These properties of a large class of iterative equations turn out to be difficult to prove, and, case in point, they were not predicted. Once aware of the period doubling behavior and its ubiquity, one can set out to understand, and eventually formally prove, why period doubling often occurs in iterative equations. Indeed, Feigenbaum universality was eventually understood in a profound way, but only *after* it was discovered numerically. This is a historical example where numerical calculations lead to an important insight. It opens our eyes to the possibility that numerical computations can be used not only to obtain mundane numbers and solutions, but also to make original discoveries and gain profound insight.

(And to elevate the role numerics played in this discovery even further, even the mathematical proof of the universality of period doubling was initially computer-assisted; it involved an inequality that was shown to hold only by numerical evaluation.)

Brainteaser: Returning to the issue of which equations can be solved analytically, it helps to be able to distinguish problems that require numerical methods from those that do not. As an exercise, can you judge which of the following can be obtained analytically, in closed form?

(i) All of the roots of $x^5 - 7x^4 + 2x^3 + 2x^2 - 7x + 1 = 0$

(ii) The integral $\int x^2/(2 + x^7)dx$

(iii) The sum $\sum_{k=1}^{N} k^4$

(iv) The solution to the difference equation $3y_{n+1} = 7y_n - 2y_{n-1}$

(v) $\exp(A)$, where A is a 2×2 matrix, $A = ((2,-1),(0,2))$, and the exponential of the matrix is defined by its power series.

EXERCISES

1.1 Survey of computing background

    a. Have you taken any courses on numerical methods or scientific computing? If yes, which course(s)?

    b. What operating system do you commonly use?
- ☐ Linux
- ☐ Mac OS
- ☐ Windows
- ☐ Other (specify):

    c. What programming languages do you know?
- ☐ None
- ☐ C
- ☐ C++
- ☐ Fortran
- ☐ Java
- ☐ Python
- ☐ R
- ☐ Other (specify):

    d. What software tools do you use?
- ☐ IDL
- ☐ Matlab or Octave
- ☐ Mathematica
- ☐ Emacs editor

☐ vi or vim editor

☐ Microsoft Excel, Google Sheets, or similar spreadsheet software

☐ IPython

☐ Gnuplot

☐ awk

☐ Perl

☐ Other (specify):

## CHAPTER 2

A Few Concepts from Numerical Analysis

A systematic treatment of numerical methods is provided in conventional courses and textbooks on numerical mathematics. Numerical analysis refers to the large portion of numerical mathematics that deals with continuous problems. A few especially common issues, which emerge in similar form in many numerical methods, are discussed here.

2.1 ROOT FINDING: FAST AND UNRELIABLE

We consider the problem of solving a single (nonlinear) equation, where a function of one variable equals a constant. Suppose a function $f(x)$ is given and we want to find its root(s) x^*, such that $f(x^*) = 0$.

A popular method is that named after Newton. The tangent at any point can be used to guess the location of the root. Since by Taylor expansion

$$f(x^*) = f(x) + f'(x)(x^* - x) + O(x^* - x)^2,$$

the root can be estimated as $x^* \approx x - f(x)/f'(x)$ when x is close to x^*. The procedure is applied repeatedly:

$$x_{n+1} = x_n - f(x_n)/f'(x_n).$$

Newton's method is so fast, computers may use it internally to evaluate the square root of a number, $x = \sqrt{a}$, with $f(x) = x^2 - a$. For example, the square root of 2 can be quickly calculated with $x_{n+1} = x_n - (x_n^2 - 2)/(2x_n) = x_n/2 + 1/x_n$. This only requires divisions and additions. With a starting value of $x_0 = 2$, we obtain the numbers shown in Table 2.1. Eight significant digits are achieved after only four iterations.

To give another example, it is possible to quickly solve $\sin(3x) = x$ in this way, by finding the roots of $f(x) = \sin(3x) - x$. Starting with $x_0 = 1$, the

TABLE 2.1 Newton method used to obtain a square root.

n	x_n
0	2
1	1.5
2	1.41666...
3	1.414216...
4	1.4142136...
$\sqrt{2}$	1.4142136...

TABLE 2.2 Newton's method applied to $\sin(3x) - x = 0$ with two different starting values.

n	x_n	
	$x_0 = 1$	$x_0 = 2$
0	1	2
1	0.7836...	3.212...
2	0.7602...	2.342...
3	0.7596...	3.719...
4	0.7596...	-5.389...

procedure produces the numbers shown in the second column of Table 2.2. The sequence quickly approaches a constant.

Newton's method may be extremely fast to converge, but it can easily fail to find a root. With $x_0 = 2$ instead of $x_0 = 1$ the iteration diverges, as apparent from the last column of Table 2.2.

Robustness may be preferable to speed. Is there a method that is certain to find a root? A simple and robust method is bisection, which follows the "divide-and-conquer" strategy. Suppose we start with two x-values where the function $f(x)$ has opposite signs. Any continuous function must have a root between these two values. We then evaluate the function halfway between the two endpoints and check whether it is positive or negative there. This restricts the root to that half of the interval on whose ends the function has opposite signs. Table 2.3 shows an example. With the bisection method the accuracy only doubles at each step, but the root is found for certain.

There are more methods for finding roots than the two just mentioned, Newton's and bisection. Each method has its advantages and disadvantages. Bisection is most general but is awfully slow. Newton's method is less general but much faster. Such a trade-off between generality and efficiency is often inevitable. This is so because efficiency is often achieved by exploiting a specific property of a system. For example, Newton's method makes use of the differentiability of the function, whereas the bisection method does not and works equally well for functions that cannot be differentiated.

The bisection method is guaranteed to succeed only if it brackets a root to begin with. There is no general method to find appropriate starting values, nor

TABLE 2.3 Bisection method applied to $\sin(3x) - x = 0$.

n	x_{lower}	x_{upper}
0	0.1	2
1	0.1	1.05
2	0.575	1.05
3	0.575	0.8125
4	0.6938...	0.8125
5	0.7531...	0.8125
⋮	⋮	⋮
15	0.7596...	0.7597...
16	0.7596...	0.7596...

do we generally know how many roots there are. For example, a function can reach zero without changing sign; our criterion for bracketing a root does not work in this case. Moreover, even a continuous function can in any interval drop rapidly, cross zero, and then increase again, making it impossible to exclude the existence of roots (Figure 2.1). Exercise 2.1 will illustrate that with a dramatic example. Unless additional information is known about the properties of the function, a search would have to explore each arbitrarily small interval to make sure it finds all and any roots. (If the nonlinear equation is a polynomial, then a great deal can be deduced about its roots analytically. For example, it is easy to find out how many roots it has. Specialized root-finding methods are available for polynomials.)

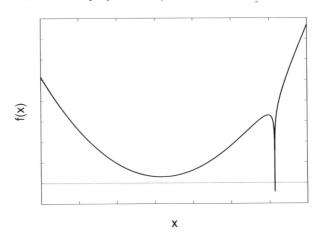

FIGURE 2.1 A continuous function with a sudden change, which may cause a numerical root finding procedure to miss the roots.

The problem becomes even more severe for finding roots in more than one variable, say under the simultaneous conditions $g(x, y) = 0, f(x, y) = 0$.

Figure 2.2 illustrates the situation. Even a continuous function could dip below zero over only a small domain, and a root searching procedure may have difficulty finding it. Not only is the space of possibilities large, but the bisection method cannot be extended to several variables.

The situation is exacerbated with more variables, as the space of possibilities is vast. The behavior of functions in ten dimensions can be difficult to trace indeed. The conclusion is clear and cold: there is no method that is guaranteed to find all roots. This is not a deficiency of the numerical methods, but it is the intrinsic nature of the problem. Unless a good, educated initial guess can be made or enough can be deduced about the solutions analytically, finding roots in more than a few variables may be fundamentally and practically impossible. This is in stark contrast to solving a system of *linear* equations, which is an easy numerical problem. Ten linear equations in ten variables can be solved numerically with great ease, except perhaps for the most pathological of coefficients, whereas numerical solution of ten nonlinear equations may remain inconclusive even after a significant computational effort. (Something analogous can be said about a purely analytical approach: a linear system can be solved analytically, whereas a system of nonlinear equations most often can not.)

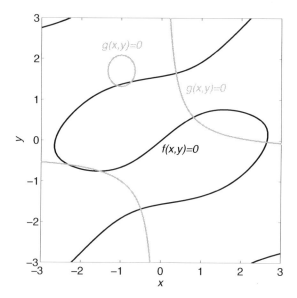

FIGURE 2.2   Roots of two functions f, g in two variables (x, y). The roots are where the contours intersect.

Root finding can be a numerically difficult problem, because there is no method that always succeeds.

2.2   ERROR PROPAGATION

Numerical problems can be difficult for other reasons too.

When small errors in the input data, of whatever origin, can lead to large errors in the resulting output data, the problem is called "numerically badly-conditioned" or if the situation is especially bad, "numerically ill-conditioned." An example is solving the system of linear equations

$$
\begin{aligned}
x - y + z &= 1 \\
-x + 3y + z &= 1 \\
y + z &= 2
\end{aligned}
$$

Suppose there is an error ϵ in one of the coefficients such that the last equation becomes $(1 + \epsilon)y + z = 2$. The solution to these equations is easily worked out as $x = 4/\epsilon$, $y = 1/\epsilon$, and $z = 1 - 1/\epsilon$. Hence, the result is extremely sensitive to the error ϵ. The reason is that for $\epsilon = 0$ the system of equations is linearly dependent: the sum of the left-hand sides of the first two equations is twice that of the third equation. The right-hand side does not follow the same superposition. Consequently the unperturbed equations ($\epsilon = 0$) have no solution. The situation can be visualized geometrically (Figure 2.3). Each of the equations describes an infinite plane in a three-dimensional space (x, y, z) and the point at which they intersect represents the solution. None of the planes are parallel to each other, but their line intersections are. For $\epsilon = 0$, the three planes do not intersect at a common point, but tilting a plane slightly (small ϵ) would lead to an intersection far out. This is a property of the problem itself, not the method used to solve it. No matter what method is utilized to determine the solution, the uncertainty in the input data will lead to an uncertainty in the output data. If a linear system of equations is linearly dependent, it is an ill-conditioned problem.

The extent to which the outcome depends on the initial errors can often be quantified with a "condition number", which measures how the output value(s) change in response to small changes in the input value(s). This condition number κ can be about the absolute or relative error, and represents the proportionality factor between input error and output error.

Computing that condition number may be computationally more expensive than obtaining the solution. For example, for a large linear system that is *almost* degenerate, one will have no problem obtaining its solution, but finding out whether this solution is robust requires an additional and larger effort.

We can investigate the condition number of the roots of a single function $f(x; a)$ that depends on a parameter a. For a root $f(x^*; a) = 0$. The error can be simply defined by absolute values $|dx^*| = \kappa|da|$. A small change in a will change f by $(\partial f/\partial a)da$. And a small change in x^* will change f by $(\partial f/\partial x)dx = f'(x)dx$. The two must cancel each other, $0 = \frac{\partial f}{\partial a}da + f'(x)dx$. Hence the solution shifts by $dx = -\frac{\partial f}{\partial a}/f'(x)da$ and the condition number (for

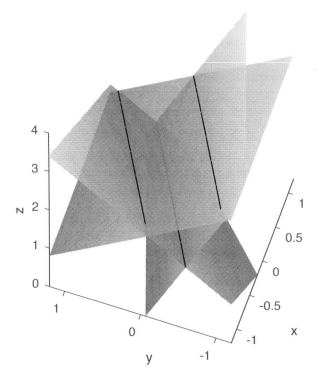

FIGURE 2.3 The planes that correspond to the degenerate linear system and their line intersections.

the absolute change of the root in response to a small perturbation in a) is

$$\kappa = \frac{dx^*}{da} = \left| \frac{\partial f / \partial a}{f'} \right|_{x=x^*}$$

If $f'(x^*) = 0$, that is, if the root is degenerate, the condition number is infinite. An infinite condition number does not mean all is lost; it only means that the solution changes faster than proportional to the input error. For example $dx \propto \sqrt{da}$ has $\kappa = \infty$. The root also becomes badly-conditioned when the numerator is large, that is, when f is highly sensitive to a. A root is well-conditioned unless one of these two conditions applies.

A special case of this is the roots of polynomials

$$p(x; a_0, \ldots, a_N) = \sum_{n=0}^{N} a_n x^n$$

At each root $p(x_j; a_0, \ldots, a_N) = 0$. For a polynomial $\partial p / \partial a_n = x^n$. Hence the condition number for the absolute change of the j-th root in response to a small perturbation of coefficient a_n is

$$\kappa_{j,n} = \left| \frac{x_j^n}{p'(x_j)} \right|$$

For a degenerate root, $p'(x_j) = 0$, κ is again infinite. For a high-degree polynomial x_j^n can be large, and if this is not compensated by a large $p'(x_j)$, then κ can be high even if the root is not degenerate. Hence, the roots of high-degree polynomials can be badly-conditioned.

Another method for determining the robustness of a numerical solution, especially when a condition number is unavailable, is the "randomization" technique, where small random values are added to each input parameter to find out whether the solution is sensitive to the changes. This is an empirical approach, but a large number of combinations may provide a useful assessment of the condition. This approach is easy to implement even for the most complex of equations. It still requires a computational effort many times that needed to calculate the solution.

2.3 NUMERICAL INSTABILITIES

The theme of error propagation has many facets. Errors introduced during the calculation, for example by roundoff, can also become critical, in particular when errors are amplified not only once, but repeatedly. One such example for the successive propagation of inaccuracies is given here.

Consider the difference equation $3y_{n+1} = 7y_n - 2y_{n-1}$ with the two starting values $y_0 = 1$ and $y_1 = 1/3$. The analytical solution to this equation is $y_n = 1/3^n$. If we iterate numerically with initial values $y_0 = 1$ and $y_1 = 0.3333$ (which approximates $1/3$), then the third column of Table 2.4 shows what happens. For comparison, the second column in the table shows the numerical value of the exact solution. The numerical iteration breaks down after a few steps.

Even if y_1 is assigned exactly $1/3$ in the computer program, using single-precision numbers, the roundoff errors spoil the solution (last column in Table 2.4). This iteration is "numerically unstable"; the numerical solution quickly grows away from the true solution.

The reason for the rapid accumulation of errors can be understood from the analytical solution of the difference equation with general initial values: $y_n = c_1(1/3)^n + c_2 2^n$, which is the sum of a decaying term and a growing term. The initial conditions for the above starting values are such that $c_1 = 1$ and $c_2 = 0$, so that the growing branch of the solution vanishes, but any error seeds the exponentially growing contribution. Indeed, the last few entries in the third column of Table 2.4 double at every iteration, because they are dominated by the 2^n contribution.

Numerical instabilities are due to the method rather than the mathematical nature of the equation being solved, contrary to the examples in the previous section 2.2, where the source of error was the coefficients in the equations

TABLE 2.4 Numerical solution of the difference equation $3y_{n+1} = 7y_n - 2y_{n-1}$ with initial error (third column) and roundoff errors (last column) compared to the exact numerical values (second column).

n	y_n		
	$y_n = 1/3^n$	$y_1 = 0.3333$	$y_1 = 1./3.$
0	1	1.	1.
1	0.333333	0.3333	0.333333
2	0.111111	0.111033	0.111111
3	0.037037	0.0368777	0.0370372
4	0.0123457	0.0120257	0.0123459
5	0.00411523	0.00347489	0.00411569
6	0.00137174	9.09399E-05	0.00137267
7	0.000457247	-0.0021044	0.000459095
8	0.000152416	-0.0049709	0.00015611
9	5.08053E-05	-0.0101958	5.81938E-05
10	1.69351E-05	-0.0204763	3.17123E-05
11	5.64503E-06	-0.0409809	3.51994E-05
12	1.88168E-06	-0.0819712	6.09904E-05
13	6.27225E-07	-0.163946	0.000118845
14	2.09075E-07	-0.327892	0.000236644

themselves. These instabilities can occur not only for difference equations, but also for differential equations (as will be described in chapter 7), and especially for partial differential equations (chapter 15), where the source of the errors is not roundoff but discretization.

———

In summary, we have encountered a number of issues that come up in numerical computations. There is commonly a tradeoff between the generality of a numerical method on one hand and efficiency on the other. There may be no algorithm that succeeds for certain, as for root finding in one variable (in principle) and root finding in many variables (in practice). The propagation of errors in input data or due to roundoff can lead to difficulties. Solutions can be highly sensitive to uncertainties in the input data, e.g., a system of linear equations that is nearly degenerate. This sensitivity can be assessed, but that involves more work than obtaining the solution. Difficulties may be intrinsic to the problem or intrinsic to the method. The themes of computational efficiency—demands on speed, memory, and data transfer—will be treated in several later chapters.

Recommended Reading: A practically oriented classic is Press, Teukolsky, Vetterling, and Flannery's *Numerical Recipes*. This voluminous book describes a broad and selective collection of methods and provides insightful lessons in numerical mathematics; see `http://numerical.recipes` for further information. Good textbooks include Burden, Faires, and Burden, *Numerical Analysis*, and the more advanced and topical Stoer & Bulirsch, *Introduction to Numerical Analysis*.

EXERCISES

2.1 a. Plot the function $f(x) = 3\pi^4 x^2 + \ln((x - \pi)^2)$ in the range 0 to 4.
    b. Prove that $f(x)$ has two roots, $f(x) = 0$.
    c. Estimate the distance between the two roots.
    d. Plot the function in the vicinity of these roots. Does the graph of the function change sign, as it must?

2.2 Show that when the Newton method is used to calculate the square root of a number $a \geq 0$, it converges for all initial conditions $x_0 > 0$.

2.3 Degenerate roots of polynomials are numerically ill-conditioned. For example, $x^2 - 2x + 1 = 0$ is a complete square $(x - 1)^2 = 0$ and has the sole degenerate root $x = 1$. Suppose there is a small error in one of the three coefficients, $(1 - \delta_2)x^2 - 2(1 + \delta_1)x + (1 - \delta_0) = 0$. Perturbing each of the three coefficients independently and infinitesimally, determine how the relative error ϵ of the root depends on the relative error of each coefficient.

2.4 Calculate the condition numbers for addition, subtraction, multiplication, and division of two numbers of the same sign and based on relative error. When the relative input errors are ϵ_1 and ϵ_2, and the relative error of the output is ϵ_{12}, then a single condition number κ can be defined as $|\epsilon_{12}| = \kappa \max(|\epsilon_1|, |\epsilon_2|)$.

2.5 Consider the linear system $A\mathbf{x} = \mathbf{b}$ where the 2×2 matrix

$$A = \begin{pmatrix} c & d \\ e & f \end{pmatrix}$$

    a. Calculate the solution $\mathbf{x}$ analytically/symbolically.
    b. Under what conditions is the solution well-conditioned, badly-conditioned, and ill-conditioned?

## CHAPTER 3

Roundoff & Number Representation

3.1 NUMBER REPRESENTATION

In a computer every real number is represented by a sequence of bits, often 64 bits (8 bytes). (1 byte is *always* 8 bits.) One bit is for the sign, and the distribution of bits for mantissa and exponent can be platform dependent. Almost universally, however, a 32-bit number will have 8 bits for the exponent and 23 bits for the mantissa (as illustrated in Figure 3.1). In the decimal system this corresponds to a maximum/minimum exponent of ± 38 and approximately 7 decimal digits. The relation between the number of binary bits and decimal accuracy can be worked out as follows. The 23 bits of the mantissa provide $2^{23} \approx 10^7$ different numbers, and therefore about 7 significant digits. The exponent can be negative or positive. Half of the 2^8 numbers can be used for the positive exponent: $2^{2^7} \approx 10^{38.5}$, so the largest number has a decimal exponent of $+38$. The 8 bits, $2^{2^8} \approx 10^{77.1}$, can represent decimal exponents from -38 to $+38$, which are 77 in number.

$$\underset{\text{sign}}{0} \ \underbrace{01011110}_{\text{exponent}} \ \underbrace{00111000100010110000010}_{\text{mantissa}}$$

$$\underset{\text{sign mantissa}}{+\ \underbrace{1.23456}} \ \underset{\text{exponent}}{\underbrace{\text{E-6}}}$$

FIGURE 3.1 Typical representation of a real number with 32 bits.

The 7 significant decimal digits are the *typical* precision. There are number ranges where the binary and decimal representations mesh well, and ranges where they mesh poorly. The number of significant digits is at least 6 and at most 9. (This can be figured out with a brute force loop through all floating-point numbers; there are only about 4 billion of them.)

For a 64-bit number (8 bytes) there are 11 bits for the exponent (which translates to decimal exponents of ± 308) and 52 bits for the mantissa, which provides around 16 decimal digits of precision. And here is why: $2^{52} \approx 10^{16}$ and $2^{2^{10}} \approx 10^{+308}$.

Single-precision numbers are typically 4 bytes long. Use of double-precision variables doubles the length of the representation to 8 bytes. On some machines there is a way to extend beyond double, up to quadruple precision (typically 128 bit). Single and double precision variables do not necessarily correspond to the same byte-length on every machine.

The mathematical constant π up to 36 significant decimal digits (usually enough for quadruple precision) is

$\leftarrow$ single $\rightarrow$
3.14159265   3589793   23846264338327950288
$\longleftarrow$ double $\longrightarrow$

Using double-precision numbers is usually barely slower than single-precision, if at all. Some processors always use their highest precision even for single-precision variables, and the extra step to convert between number representations makes single-precision calculations actually slower. Double-precision numbers do, however, take twice as much memory.

Several general-purpose math packages offer arbitrary-precision arithmetic, should this ever be needed. Computationally, arbitrary-precision calculations are disproportionally slow, because they have to be emulated on a software level, whereas single- and double-precision floating-point operations are hardwired into the computer's central processor.

Many fractions have infinitely many digits in decimal representation, e.g.,

$$\frac{1}{6} = 0.1666666\ldots$$

The same is true for binary numbers; only that the exactly represented fractions are fewer. The decimal number 0.5 can be represented exactly as 0.100000..., but decimal 0.2 is in binary form

0.00110011001100110...

and hence not exactly representable with a finite number of digits. This causes a *truncation error*. In particular, decimals like 0.1 or 10^{-3} have an infinitely long binary representation. (The fact that binary cannot represent arbitrary decimal fractions exactly is a nuisance for accounting software, where everything needs to match to the cent.) For example, if a value of 9.5 is assigned it will be 9.5 exactly, but 9.1 carries a representation error. One can see this by using the following Python commands, which print numbers to 17 digits after the comma.

```
>>> print '%.17f' % 9.5
9.50000000000000000
>>> print '%.17f' % 9.1
9.09999999999999964
```

(The first % symbol indicates that the following is the output format. The second % symbol separates this from the number.) For the same reason, $0.1 + 0.1 + 0.1 - 0.3$ is not zero,

```
>>> 0.1+0.1+0.1-0.3
5.551115123125783e-17
```

A discrepancy of this order of magnitude, $\approx 10^{-16}$, incidentally makes clear that by default Python, or at least the implementation used here, represents floating-point numbers as 8-byte variables. Any if condition for a floating-point number needs to include a tolerance, e.g., not if (a==0.8), but if (abs(a-0.8)<1e-12), where 10^{-12} is an empirical choice that got to be safely above a (relative) accuracy of 10^{-16}, for whatever range a takes on in this program.

In terms of more formal mathematics, in the language of algebraic structures, the set of floating-point numbers ($\mathbb{F}$) does not obey the same rules as the set of real numbers ($\mathbb{R}$). For example, $(a + b) + c$ may be different from $a + (b + c)$. The associative property, that the order in which operations are performed does not matter, does not necessarily hold for floating-point numbers. For example,

```
>>> (1.2-1)-0.2
-5.551115123125783e-17
>>> 1.2-(1+0.2)
0.0
```

Some programming languages assume that, unless otherwise instructed, all numbers are double-precision numbers. Others look at the decimal point, so that 5 is an integer, but 5. is a floating-point number. Often that makes no difference in a calculation at all, but sometimes it makes a crucial difference. For example, for an integer division 4/5=0, whereas for a floating-point division 4./5.=0.8. This is one of those notorious situations where the absence of a single period can introduce a fatal bug in a program. Even 4./5 and 4/5. will yield 0.8. For this reason some programmers categorically add periods after integers that are meant to be treated as floating-point numbers, even if they are redundant; that habit helps to avoid this error.

When a number is represented with a fixed number of bits, there is necessarily a maximum and minimum representable number; exceeding them means an "overflow" or "underflow." This applies to floating-point numbers as well as to integers. For floating-point numbers we have already determined these limits. Currently the most common integer length is 4 bytes. Since 1 byte is 8 bits, that provides $2^{4 \times 8} = 2^{32} \approx 4 \times 10^9$ different integers. The C language and numerous other languages also have *unsigned* integers, so all bits can be used for positive integers, whereas the regular 4-byte integer goes from about -2×10^9 to about $+2 \times 10^9$. It is prudent not to use loop counters that go beyond 2×10^9.

Some very high-level programming languages, such as Python and Mathematica, use *variable* byte lengths for integers, in contrast to the fixed byte length representations we have discussed so far. If we type 10**1000 at a Python prompt, which would overflow a 4-byte integer, and even overflow the maximum exponent of an 8-byte floating-point, then it automatically switches to a longer integer representation.

The output of 10**1000 in Python 2 will be appended with the letter L, which indicates a "long" integer, that would not fit within an integer of regular byte length. Similarly, floating-point numbers are sometimes written as 1e6 or 1d6 to distinguish single-precision from double-precision numbers. The numbers 1e6 and 1d6 are exactly the same, but 1e-6 and 1d-6 are not, because of truncation error. For example, in Fortran the assignment a=0.1 is less accurate than a=0.1d0, even when a is declared as a double-precision variable in both cases. Letters that indicate the variable type within a number, such as e, d, and L, are known as "literals". Only some languages use them.

3.2 IEEE STANDARDIZATION

The computer arithmetic of floating-point numbers is defined by the IEEE 754 standard (originally 754-1985, then revised by the 854-1987 and 754-2008). It standardizes number representation, roundoff behavior, and exception handling. All three components are will be described in this chapter.

TABLE 3.1 Specifications for number representation according to the IEEE 754 standard.

	single	double
bytes	4	8
bits for mantissa	23	52
bits for exponent	8	11
significant decimals	6–9	15–17
maximum finite	3.4E38	1.8E308
minimum normal	1.2E-38	2.2E-308
minimum subnormal	1.4E-45	4.9E-324

Table 3.1 summarizes the IEEE standardized number representations, partly repeating what is described above. What matters for the user are the number of significant decimals, and the maximum and minimum representable exponent. When the smallest (most negative) exponent is reached, the mantissa can be gradually filled with zeros, allowing for even smaller numbers to be represented, albeit at less precision. Underflow is hence gradual. These numbers are referred to as "subnormals" in Table 3.1.

As a curiosity, $\tan(\pi/2)$ does not overflow with standard IEEE 754 numbers, neither in single nor double precision. This is straightforward to demonstrate. If π is not already defined intrinsically, assign it with enough digits, given above, divide by two, and take the tangent; the result will be finite. Or, if π is already defined, type

```
>>> tan(pi/2)
1.633123935319537e+16
```

In fact the tangent does not overflow for any argument.

As part of the IEEE 754 standard, a few bit patterns have special meaning and serve as "exceptions". There is a bit pattern for numbers exceeding the maximum representable number: Inf (infinity). There are also bit patterns for -Inf and NaN (not a number). For example, 1./0. will produce Inf. An overflow is also an Inf. There is a positive and a negative zero. If a zero is produced as an underflow of a tiny negative number it will be $-0.$, and $1./(-0.)$ produces -Inf. A NaN is produced by expressions like $0./0.$, $\sqrt{-2.}$, or Inf-Inf. Exceptions are intended to propagate through the calculation, without need for any exceptional control, and can turn into well-defined results in subsequent operations, as in 1./Inf or in if (2.<Inf). If a program aborts due to exceptions in floating-point arithmetic, which can be a nuisance, it does not comply with the standard. IEEE 754 floating-point arithmetic is algebraically complete; every algebraic operation produces a well-defined result.

Roundoff under the IEEE 754 standard is as good as it can be for a given precision. The standard requires that the result must be as if it was first computed with infinite precision, and then rounded. This is a terrific accomplishment; we get numbers that are literally accurate to the last bit. The error never exceeds half the gap of the two machine-representable numbers closest to the exact result. (There are actually several available rounding modes, but, for good reasons, this is the default rounding mode.) This applies to the elementary operations $(+, -, /, \times)$ as well as to the remainder (in many programming languages denoted by %) and the square root $\sqrt{}$. Halfway cases are rounded to the nearest even (0 at the end) binary number, rather than always up or always down, because rounding in the same direction would be more likely to introduce a statistical bias, as minuscule as it may be.

The IEEE 754 standard for floating-point arithmetic represents the as-good-as-it-can-be case. But to what extent are programming platforms compliant with the standard? The number representation, that is, the partitioning of the bits, is nowadays essentially universally implemented on platforms one would use for scientific computing. That means for an 8-byte number, relative accuracy is about 10^{-16} and the maximum exponent is $+308$ nearly always. Roundoff behavior and exception handling are often available as options, because obeying the standard rigorously comes with a penalty on speed. Compilers for most languages provide the option to enable or disable the roundoff and exception behavior of this IEEE standard. Certainly for C and Fortran, ideal rounding and rigorous handling of exceptions can be enforced on most machines. Many general-purpose computing environments also comply with the IEEE 754 standard. Pure Python stops upon division by zero—which is a violation of the standard, but the NumPy module is IEEE compliant.

Using exactly representable numbers allows us to do calculations that incur no roundoff at all, at least when IEEE 754 is enabled. Of course every integer, even when defined as a floating-point number, is exactly representable. For example, addition of 1 or multiplication by 2 does not incur any roundoff at all. Normalizing a number to avoid an overflow is better done by dividing by a power of 2 than by a power of 10, due to truncation error. Factorials can be calculated, without loss of precision, using floating-point numbers; when the result is smaller than $\sim 2 \times 10^9$ it will provide the exact same answer as a calculation with integers would have, and if it is larger it will be imprecise, but at least it will not overflow as integers would.

NaNs as floating point numbers can be tricky. For example,

```
if (x >= 0) { printf("x is positive or zero\n"); }
    else { printf("x is negative\n"); }
```

would identify x=NaN as negative, although it is not. It is more robust to re-place the `else` statement with `{if (x<0) printf("x is negative\n"); }`. The result of a comparison can only be true or false (a boolean/logical vari-able cannot have a NaN value), and all comparisons with NaN return false. Whereas the result of NaNs in elementary operations is clear (if one argument is NaN, then the result is NaN), its impact upon other functions can be am-biguous. The meaning of NaN (Not a Number or Not any Number) can be twofold. One is as a number of unknown value; the other is as a missing value. The result of `max(1,2,NaN)` should be `NaN` in the former case and 2 in the latter.

To summarize part of the IEEE 754 number representation standard: there are three special values: NaN, +Inf, and −Inf, and two zeros, +0 and −0. It has signed infinities and signed zeros. It allows for five exceptions: invalid operation (produces NaN), overflow (produces +/-Inf), underflow (produces +/-0), division by zero (produces +/−Inf), and inexact. The last of these exceptions, which had not been mentioned yet, tells us whether any rounding has occurred at all, a case so common that no further attention is usually paid to it. IEEE 754 applies to floating point numbers, and not to other types of variable; so if an integer overflows, we may really be in trouble.

The rigor with which roundoff is treated is wonderfully illustrated with the following example: The numerical example of a chaotic iteration in chapter 1 is extremely sensitive to the initial condition and subsequent roundoff. Never-theless, these numbers, even after one thousand iterations, can be reproduced *exactly* on a different computer and with a different programming language (Exercise 3.1). Of course, given the sensitivity to the initial value, the result is quantitatively incorrect on all computers; after many iterations it is entirely different from a calculation using infinitely many digits.

3.3 ROUNDOFF SENSITIVITY

Using the rules of error propagation, or common sense, we recognize situations that are sensitive to roundoff. If x and y are real numbers of the same sign, their sum $x + y$ has an absolute error that adds the two individual absolute errors, and the relative error is at most as large as the relative error of x or y. Hence, adding them is insensitive to roundoff. On the other hand, $x - y$ has increased relative error (Exercise 2.4). The relative error of a product of two numbers is the sum of the relative errors of its factors, so multiplication is also not roundoff sensitive. The relative error of a ratio of two numbers is also the sum of relative errors, so division is not sensitive to roundoff either. To go through at least one of these error propagations:

$$\frac{x}{y}(1 + \epsilon_{x/y}) = \frac{x(1 + \epsilon_x)}{y(1 + \epsilon_y)} \approx \frac{x}{y}(1 + \epsilon_x)(1 - \epsilon_y) \approx \frac{x}{y}(1 + \epsilon_x - \epsilon_y)$$

and therefore the relative error of the ratio is $|\epsilon_{x/y}| \leq |\epsilon_x| + |\epsilon_y|$. For divisions, we only need to worry about overflows or underflows, in particular division by zero. Among the four elementary operations only subtraction of numbers of equal sign or addition of numbers of opposite sign increase the relative error.

An instructive example is solving a quadratic equation $ax^2 + bx + c = 0$ numerically. In the familiar solution formula

$$x = \frac{-b \pm \sqrt{b^2 - 4ac}}{2a}$$

a cancellation effect will occur for one of the two solutions if ac is small compared to b^2. The remedy is to compute the smaller root from the larger. For a quadratic polynomial the product of its two roots equals $x_1 x_2 = c/a$, because $ax^2 + bx + c = a(x - x_1)(x - x_2)$. If b is positive then one solution is obtained by the equation above, $x_1 = -q/(2a)$, with $q = b + \sqrt{b^2 - 4ac}$, but the other solution is obtained as $x_2 = c/(ax_1) = -2c/q$. This implementation of the solution of quadratic equations requires no extra line of code; the common term q could be calculated only once and stored in a temporary variable, and the sign of b can be accommodated by using the sign function $\mathrm{sgn}(b)$, $q = b + \mathrm{sgn}(b)\sqrt{b^2 - 4ac}$. To perfect it, a factor of $-1/2$ can be absorbed into q to save a couple of floating-point operations.

```
!straight-forward version  ! roundoff-robust version
d = sqrt(b**2-4*a*c)       q=-(b+sgn(b)*sqrt(b**2-4*a*c))/2
x1 = (-b + d)/(2*a)        x1 = q/a
x2 = (-b - d)/(2*a)        x2 = c/q
```

We usually do not need to bother writing an additional line to check whether a is zero. If a division by zero does occur, a modern computer will either complain or it is properly taken care of by the IEEE standard, which would produce an `Inf` and continue with the calculation in a consistent way.

Sometimes an expression can be recast to avoid cancellations that lead to increased sensitivity to roundoff. For example, $\sqrt{1+x^2}-1$ leads to cancellations when x is close to zero, but the equivalent expression $x^2/(\sqrt{1+x^2}+1)$ has no such problem. A basic example of an alternating series whose cancellation error can be avoided is

$$1 - \frac{1}{2} + \frac{1}{3} - \frac{1}{4} + -... = \frac{1}{1 \cdot 2} + \frac{1}{3 \cdot 4} + ...$$

The version to the left involves cancellations, the one to the right does not. There is no need to evaluate this infinite series numerically—it sums up to $\ln 2$ —, but it illustrates the concept.

An example of unavoidable cancellations is finite-difference formulae, like $f(x+h)-f(x)$, where the value of a function at point x is subtracted from the value of a function at a nearby point $x+h$. (An illustration of the combined effect of discretization and roundoff errors in a finite-difference expression will be given in Figure 6.1.)

A potential problem also arises for $\mathrm{acos}(x)$ and $\mathrm{asin}(x)$, where the argument has to be between -1 and $+1$. If x is slightly above 1 due to roundoff, it will lead to a NaN. An extra line may be necessary to prevent that. (It happens so that IEEE 754 guarantees that $-1 \le x/\sqrt{x^2+y^2} \le 1$, unless the denominator underflows.)

We conclude this section by applying our knowledge of roundoff to the problem of calculating the distance on a sphere. The great circle distance $\Delta\sigma$ on a sphere is

$$\cos(\Delta\sigma) = \sin\phi_1 \sin\phi_2 + \cos\phi_1 \cos\phi_2 \cos(\Delta\lambda)$$

which follows from one or the other law of spherical trigonometry. Here, $\Delta\sigma$ is the angular distance along a great circle, ϕ_1 and ϕ_2 are the latitudes of the two points, and $\Delta\lambda = \lambda_2 - \lambda_1$ is the difference in longitudes. This expression is prone to rounding errors when the distance is small. This is clear from the left-hand side alone. For $\Delta\sigma \approx 0$, the cosine is almost 1. Since the Taylor expansion of the cosine begins with $1 - \Delta\sigma^2/2$, the result will be dominated by the truncation error at $\Delta\sigma^2/2 \approx 10^{-16}$ for 8-byte floating-point numbers. The volumetric mean radius of the Earth is 6371 km, so the absolute distance comes out to $2\pi \times 6371 \times 10^3 \times \sqrt{2 \times 10^{-16}} \approx 0.6$ m. At distances much larger than that, the formula given above does not have significant rounding errors. For small distances we could use the Pythagorean relation $\Delta\sigma = \sqrt{\Delta\phi^2 + \Delta\lambda^2 \cos^2\phi}$. The disadvantage of this approach is that there will be a small discontinuity when switching between the roundoff-sensitive exact equation and the roundoff-insensitive approximate equation. Alternatively the equation can be reformulated as

$$\sin^2\left(\frac{\Delta\sigma}{2}\right) = \sin^2\left(\frac{\Delta\phi}{2}\right) + \cos\phi_1 \cos\phi_2 \sin^2\left(\frac{\Delta\lambda}{2}\right)$$

This no longer has a problem for nearby points, because $\sin(\Delta\sigma/2) \approx \Delta\sigma/2$, so the left-hand side is small for small $\Delta\sigma$, and there is no cancellation between the two terms on the right-hand side, because they are both positive. This expression becomes roundoff sensitive when the sine is nearly 1 though, that is, for nearly antipodal points. Someone has found a formula that is roundoff insensitive in neither situation, which involves more terms. The lesson from this example is that *if* 16 digits are not enough, the problem can often be fixed mathematically.

Interval arithmetic. Generally, we want rounding to the nearest number, but other rounding modes are available: round always up or always down. An advanced technique called "interval arithmetic" takes advantage of these directed roundings. Every result is represented not by one value of unknown accuracy, but by two that straddle the exact result. An upper and a lower bound are determined at every step of the calculation. Although the interval may vastly overestimate the actual uncertainty, it provides mathematically rigorous bounds. Interval arithmetic can sometimes turn a numerical calculation into a mathematically rigorous result.

Recommended Reading: The "father" of the IEEE 754 standard, William Kahan, posts roundoff-related notes online at `http://people.eecs.berkeley.edu/~wkahan/`. The links include a plain-language description of the standard, `http://people.eecs.berkeley.edu/~wkahan/ieee754status/IEEE754.PDF`. A technical summary is provided by David Goldberg, *What every computer scientist should know about floating point arithmetic*. The document is readily available online, for example at `https://docs.oracle.com/cd/E19957-01/806-3568/ncg_goldberg.html`.

EXERCISES

3.1 The iteration $x_{n+1} = 4x_n(1 - x_n)$ is extremely sensitive to the initial value as well as to roundoff. Yet, thanks to the IEEE 754 standard, it is possible to reproduce the exact same sequence on many platforms. Calculate the first 1010 iterations with initial conditions $x_0 = 0.8$ and $x_0 = 0.8 + 10^{-15}$. Use double (8-byte) precision. Submit the program as well as the values $x_{1000} \ldots x_{1009}$ to 9 digits after the decimal point. We will compare the results in class.

3.2 The expression for gravitational acceleration is GMr_i/r^3, where G is the gravitational constant 6.67×10^{-11} m^3/kg s, M is the mass of the sun 2×10^{30} kg, $r = 150 \times 10^9$ m is the distance between the Sun and Earth, and $i = 1, 2, 3$ indexes the three spatial directions. We cannot anticipate in which order the computer will carry out these floating-point operations. What are the worst possible minimum and maximum exponents of an intermediate result? Could 4-byte IEEE or 8-byte IEEE floating-point numbers overflow or underflow?

3.3 *hypot function:* The hypotenuse of a triangle is given by

$$c = \sqrt{x^2 + y^2}$$

A problem with this expression is that $x^2 + y^2$ might overflow or underflow, even when c does not. Find a way to calculate c that circumvents this problem.

3.4 The following formulae may incur large roundoff errors: a) $x - \sqrt{x}$, and b) $\cos^2 x - \sin^2 x$. Identify values of x for which they are sensitive to roundoff, and suggest an alternative formula for each which avoids this problem.

## CHAPTER 4

Programming Languages & Tools

Almost any programming language one is familiar with can be used for computational work. Some people are convinced that their own favorite programming language is superior to all other languages, and this may be because the versatility of a language increases with one's familiarity with it. Figure 4.1 categorizes programming languages from the perspective of scientific computing. They are grouped into i) high-level languages, ii) general-purpose interactive computational environments, iii) text processing utilities and scripting languages, and iv) special-purpose languages. This classification is by no means rigorous, but a practical grouping.

4.1 HIGH-LEVEL PROGRAMMING LANGUAGES

High-level programming languages received their name because they spare the programmer many details of what is happening in the CPU (Central Processing Unit), which is described by low-level machine code. From a modern point of view, "high level" languages have become mid-level languages, but that term is not used. Instead they are flanked by *"very* high-level languages." So, high-level is in contrast to very high-level, and requires relatively *detailed* instructions. High-level languages include C, C++, Fortran, and Java.

C is the common tongue of programmers and computer scientists. It is perhaps the most widely used programming language of all time, and, needless to say, there exist large program repositories. Its syntax is widely known and often imitated by other languages. Modern C includes intrinsic complex arithmetic, which was absent from early C standards. C++ is a much larger and more complex language than C. And C is a subset of C++, so every C program is also valid C++ code. C++ becomes advantageous over C when code gets large and needs to be maintained.

Fortran is a programming language tailored to the needs of scientists and

I. *High-level Programming Languages*

    C

    Fortran (77 & modern)

    C++

    Java

II. *Interactive Computational Environments (Very High-level Languages)*

    IDL (Interactive Data Language) and GDL

    Matlab and Octave

    Mathematica (Wolfram Language)

    Python

IIb. *Data Visualization & Graphing Software*

    Gnuplot

    Origin

    MatPlotLib (Python based)

III. *Text Processing Utilities & Scripting Languages*

    sed

    awk

    Perl

IV. *Special Purpose Software/Languages*

    Statistics: R, SAS, SPSS

    Laboratory: LabVIEW

    Database: SQL, Hadoop

FIGURE 4.1 Selected programming languages and tools organized from the perspective of scientific computing

engineers and as such it continues to be particularly well suited for this purpose. It is the oldest programming language that is still widely used in contemporary scientific research, and that for good reasons. There is extensive heritage, that is, code for scientific modeling. Fortran was invented in the 1950s, but underwent significant revision and expansion in the 1990s. Modern Fortran (Fortran 90 and later versions) greatly extends the capabilities of earlier Fortran standards. Fortran 77, the language standard that preceded Fortran 90, lacks dynamic memory allocation, that is, the size of an array cannot be changed while the program is executing. It also used a "fixed format" for the source code, with required blank columns at the front of each line and maximum line lengths, as opposed to the "free format" Modern Fortran and nearly all modern languages use. A major advantage of Fortran is its parallel computing abilities.

Java excels in portability. It typically does not reach the computational performance of languages like C and Fortran, but comes close to it, and in rare circumstances exceeds it.

Table 4.1 shows a program in C and in Fortran that demonstrates similarities between the two languages. The analogies are mostly self-explanatory. float/real declare single-precision floating-point variables. In C, statements need to be ended with a semicolon; a semicolon is also used to pack two Fortran commands into one line. In C, i++ is shorthand for i=i+1, and the way for-loops work is: for(starting point; end condition; increment expression) body. Array indices begin by default with 0 for C and with 1 for Fortran. C does not have a special syntax for powers, so the pow function is used that works for integer and non-integer powers alike. For print statements in C, \n adds a newline, whereas in Fortran output lines automatically end with a line break. The indents are merely for ease of reading and do not change the functionality of the programs. Plain C has few commands, but additional functionality is provided through the #include commands. For example, #include <math.h> provides basic mathematical functions, such as pow and sin. C is case-sensitive, whereas Fortran is not.

TABLE 4.1 Program examples that demonstrate similarities between two high-level languages.

```
/* C program example */        ! Fortran program example
#include <math.h>
#include <stdio.h>
void main()                    program demo
{                                  implicit none
  int i;                           integer i
  const int N=64;                  integer,parameter :: N=64
  float b,a[N];                    real b,a(N)
  b=-2.;                           b=-2.
  for(i=0;i<N;i++) {               do i=1,N
    a[i]=sin(i/2.);                  a(i)=sin((i-1)/2.)
    if (a[i]>b) b=a[i];             if (a(i)>b) b=a(i)
  }                                end do
  b=pow(b,5.); b=b/N;              b=b**5; b=b/N
  printf("%f\n",b);               print *,b
}                              end
```

Code is typically written in a text editor. Many general-purpose text editors, such as *Emacs*, *vim*, and *BBEdit*, support programming with auto-indent, syntax coloring, highlighting of matching pairs of braces, and other helpful features. Some language implementations come with their own editors (as part of their "integrated development environment", but many programmers prefer the universally useful general-purpose text editors. The source code is then compiled. For example, on a command prompt gcc example.c produces an executable. Type a.out, the default name of the executable, and it will output 0.015430.

Typical for high-level programming languages, in contrast to very high-level languages, is that all variables must be declared, instructions are relatively detailed, and the code needs to be explicitly compiled or built to generate an executable.

The free GNU compilers have been a cornerstone in the world of programming. They are available for many languages, including C (gcc), C++ (g++), Fortran (gfortran), and Java (gcj). Although they emerged from the Unix world, they are available in some form for all major operating systems (Windows, Mac/Unix, Android). And they are not the only free compilers out there. In addition, commercial compilers are available. These often take advantage of new hardware features more quickly than free compilers.

The role of language standards is to make sure all compilers understand the same language, although individual compilers may process nonstandard language extensions. The use of nonstandard extensions will make code less portable. Nevertheless, some nonstandard syntax is widely used. For example, Fortran's real*8, which declares an 8-byte floating-point number, has been understood by Fortran compilers since the early days of the language, but is strictly speaking not part of the language standard.

4.2 INTERACTIVE COMPUTATIONAL ENVIRONMENTS

General-purpose computational environments are the appropriate choice for most every-day computational work. Many tasks that would otherwise require lengthy programs can be accomplished with a few keystrokes.

For instance, it only takes one command to find a root. In Mathematica notation,

$$\texttt{FindRoot[sin(3 x)==x,\{x,1\}]}$$

The command searches for a solution to the equation beginning with $x = 1$. Its output is $\{x \to 0.759621\}$ (that again is Mathematica notation), which agrees with the result we obtained in Table 2.2 for the same equation.

Inverting a matrix A reduces to Inverse[A] (in Mathematica) or 1/A (in Matlab). The Mathematica command Inverse[{{1,2},{1,4}}] produces {{2,-1},{-1/2, 1/2}}, so it actually inverts the matrix symbolically. Adding a decimal point makes it clear that only a numerical answer is sought: Inverse[{{1.,2},{1,4}}] produces {{2,-1},{-0.5,0.5}}.

These tools can be used interactively, but they are essentially programming languages as well. A sequence of commands forms a program. They are *very* high-level programming languages.

Table 4.2 shows the example program above in two popular applications. These programs are considerably shorter than those in Table 4.1 and do not require variables to be declared. In Matlab an entire array of integers is created with [0:N-1] and the sine command is automatically applied to all elements in the array.

TABLE 4.2 Matlab and IDL program examples; compare to Table 4.1. In Matlab every line produces output; a semicolon at the end of the line suppresses its output, so the absence of a semicolon in the last line implies that the result is written to the screen.

```
% Matlab program example      ; IDL program example
N=64;                         N=64
i=[0:N-1];                    a=FLTARR(N)
a=sin(i/2);                   FOR i=0,N-1 DO a(i)=sin(i/2)
b=max(a);                     b=MAX(a)
b^5/N                         PRINT, b^5/N
```

Python is widely used for scientific programming. It is versatile and forgiving, and enables programmers to write code in less time. Table 4.3 is a Python implementation of the program example shown in Tables 4.1 and 4.2. Python uses whitespace indentation to delimit code blocks (instead of curly braces }, as in C, or keywords, as end in Fortran), so in the example the loop ends when the indentation ends.

TABLE 4.3 Python version of the program in Tables 4.1 and 4.2.

```
# Python program example
import numpy
N=64
a=numpy.zeros(N)
for i in range (0,N):
    a[i]=numpy.sin(i/2.)

b=max(a)
print b**5/N
```

Plain Python does not have arrays, but NumPy (Numeric Python) does. This module is loaded with the import command. NumPy also provides the mathematical functions that otherwise would need to be imported into plain Python with import math. The range command in the for loop, range(0,N), goes only from $0, \ldots, N - 1$, and not from $0, \ldots, N$.

These Python commands can be executed interactively, line by line, or they can be placed in a file and executed collectively like program code. We can start Python and type in any Python command to get output immediately. Or we can write a sequence of commands in a file, called something like `example.py`, and then feed that to Python (on a command prompt with `python example.py`).

It is amusing that in the five language examples above (C, Fortran, Matlab, IDL, and Python), each uses a different symbol to indicate a comment line: Double slash `//` or `/* ... */` in C, exclamation mark `!` or `c` in Fortran, percent symbol `%` in Matlab, semicolon `;` in IDL, and hash `#` in Python.

General-purpose computational environments that are currently widely used include: *Mathematica* began as a symbolic computation package, which still is its comparative strength. Its syntax is called the Wolfram Language. *Matlab* is particularly strong in linear algebra tasks; its name is an abbreviation of Matrix Laboratory. *Octave* is open-source software that mimics *Matlab*. *IDL (Interactive Data Language)* began as a visualization and data analysis package. Its syntax defines the Gnu Data Language (GDL). All of these software packages offer a wide range of numerical and graphical capabilities, and some of them are also capable of symbolic computations. Python and R can also function as general-purpose interactive computational environments.

4.3 RELEVANT LANGUAGE AND IMPLEMENTATION FEATURES

Interactive-mode versus built languages. Program statements may be used interactively or, alternatively, the source code may have to be built before it can be executed. The languages described in section 4.1 are built (compiled and linked). The general-purpose computational environments described in section 4.2 can each be used both ways.

Interpreted versus compiled languages. Program code can be made executable by interpreters, compilers, or a combination of both. Interpreters read and immediately execute the source program line by line. Compilers process the entire program before it is executed, which permits better checking and speed optimization. Languages that can be compiled hence execute much faster than interpreted ones. Sometimes compilation occurs in the background. Compilation and interpretation is not a property of the language itself; it is the language implementation that compiles or interprets, so there can be compiled and interpreted version of the same language. However, C, Fortran, and C++ codes are essentially always compiled, and they are commonly referred to as "compiled languages".

Common implementations of Java and Python involve a number of technical complexities that go under the names of "Virtual Machines", "bytecode", and "Just-in-time compilers". Virtual Machines are an extra layer between the program and the central processor (CPU), which emulates the underlying hardware. As a result programs become perfectly platform independent (portable). The code that runs on a Virtual Machine is called bytecode. The translation process is now split into two parts: one from the source code to bytecode and then from bytecode to machine code. Even if one of those two steps is a compilation, the other might not be. The bottom line is that Java and Python are not fully-compiled languages, and they rarely achieve the computational performance of C and Fortran.

Another, rarer possibility is source-to-source translators that convert the language statements into those of another. An example is *Cython*, which translates Python-like code to C. (Cython is not the same as CPython. The latter refers to a Python implementation written in C.)

When source code is spread over multiple files, compilation produces intermediate files, with extensions such as .o ("object file") or .pyc (Python bytecode). Unless the source code changes, these can be reused for the next built, so this saves on compilation time.

Execution speed. Python is slow. That is an overgeneralized statement, but it captures the practical situation. A factor often quoted is 20 compared to Fortran, although that is undoubtedly even more of an overgeneralization. The reasons for its inherent inefficiency have to do with how it manages memory and that it is not fully compiled. This in turn simply relates to the fact that it is a very high-level language. There are ways to improve its computational performance, and language extensions and variants such as NumPy, Cython, PyPy, and Numba help in this process, but ultimately Python was not designed for computational performance; it was designed for coding, and it excels in that. The appropriate saying is that

"Python for comfort, Fortran for speed."

Very high-level languages often do not achieve the speed possible with languages like Fortran or C. One reason for that is the trade-off between universality and efficiency—a general method is not going to be the fastest. Convenience comes at the cost of computational efficiency and extraneous memory consumption. This is formally also known as "abstraction penalty".

The bottom line is simple and makes perfect sense: higher level languages are slower in execution than lower level languages, but easier and faster to code. In a world where computing hardware is extremely fast, the former often ends up to be the better time investment.

Intrinsic functions. An arsenal of intrinsic functions is available that covers commonplace and special-purpose needs. Available mathematical functions include powers, logarithms, trigonometric functions, etc. They also most often include `atan2(y,x)`, which is the arctangent function with two instead of one argument `atan(y/x)`, which resolves the sign ambiguity of the results for the different quadrants. Even the Error function `erf` is often available. And sometimes trigonometric functions with arguments in degree are available, `sind`, `cosd`, and `tand`. In some languages an extra line of code is required to include them, such as `#include <math.h>` in C or `import math` in Python (`import numpy` will include them too; and `import scipy` will, more or less, include NumPy plus other functionality).

The symbol for powers is ^ or **. (C has no dedicated symbol). The logical *and* operator often is & or &&, and the *or* operator | or ||. The use of == as an equality condition is rather universal, and the very few languages that use a single =, use another symbol, such as :=, for a variable assignment. In contrast, the symbol for "not equal" takes, depending on the language, various forms: !=, /=, ~=. As mentioned in section 3.1, the division symbol / can mean something different for integers and floating-point numbers. In Python, // is dedicated to integer division, and called "floor division".

Sometimes the name of a function depends on the number type. For example, in C and C++ fabs is the absolute value of a double-precision floating-point number, fabsf is for a single-precision number, abs only for an integer, and cabs for a complex number. Modern C compiler implementations may be forgiving about this, and in very high-level languages it is abs for all input types.

Another feature relevant for scientific computing is *vectorized, index-free,* or *slice notation.* Fortran, Matlab, and Python are examples of languages that allow it. An operation is applied to every element of an entire array, for example A(:)=B(:)+1 adds 1 to every element. Even A=B+1 does so when both arrays have the same shape and length. This not only saves the work of writing a loop command, but also makes it obvious (to a compiler) that the task can be parallelized. With explicit loops the compiler first needs to determine whether or not the commands within the loop are independent from one another, whereas index-free notation makes it clear that there is no interdependence.

Initializations. In many languages, scalar variables are initialized with zero. Not so in Fortran, where no time is lost on potentially unnecessary initializations. Neither C nor Fortran automatically initialize arrays with zeros, so their content is whatever happens to reside in these memory locations. Some very high-level numerical environments require initializations when the array is created.

High-level versus very high-level languages. This distinction has already been dwelled upon, and there is a continuum of languages with various combinations of properties. The programming language Go or Golang, sort of a C-family language, is fully compiled but does not require variables to be declared, so it falls somewhere in the middle of this categorization. The very high-level languages are also those that are suited for interactive usage.

To again contrast very high-level with high-level languages, suppose we want to read a two-column data file such as

```
1990 14.7
2000 N/A
2010 16.2
```

where the 'N/A' stands for non-applicable (or not-available) and indicates a missing data point. This is an alphabetical symbol not a number, so it potentially spells trouble if the program expects only numbers. The following lines of C code read numbers from a file:

```
fin = fopen("file.dat","r");
for(i=0; i<3; i++) {
  fscanf(fin,"%d %f\n",&year,&T);
  printf("%d %f\n",year,T);
}
```

The output will only be the following two lines:

```
1990 14.700000
2000 14.700000
```

It terminates on the N/A and outputs the previously read entry for T, so it is doubly problematic. To remedy this problem, we would have to either read all entries as character strings and convert them to numbers inside the program, or, before even reading the data with C, replace N/A in the file with a number outside the valid range; for example if the second column is temperature in Celsius, the value -999 will do. On the other hand, with the very high-level language Matlab or Octave

```
> [year,T]=textread('file.dat','%d %f')
year =
    1990
    2000
    2010
T =
   14.700
      NaN
   16.200
```

This single line of code executes without complaint and assigns a NaN to the floating-point number T(2). (Chapter 3 has explained that NaN is part of the floating-point number representation.) This illustrates two common differences between high-level and very high-level languages: the latter are faster to code and generally more forgiving.

Cross language compilation. Programming languages can also be mixed. For example C and Fortran programs can be combined. There are technicalities that have to be observed to make sure the variables are passed correctly and have the same meaning in both programs, but language implementations are designed to make this possible. Combining Python with C provides the better of both worlds, and Python is designed to make that easily possible. Using Python as a wrapper for lower level functions can be an efficient approach. Scripts are another way to integrate executables from various languages, a topic we will turn to in chapter 14.

Choosing a programming language. We have not yet discussed groups III and IV in Figure 4.1. Group III, text processing utilities and scripting languages, is another set of very high-level languages, but with a different purpose. They are immensely useful, for example, for data reformatting and processing and they will be introduced in chapters 13 and 14. Group IV are special-purpose languages, and the short list shown in the figure ought to be augmented with countless discipline-specific and task-specific software. LabVIEW deserves special mention, because it is a *graphical* programming language, where one selects and connects graphical icons to build an instrument control or data acquisition program. It is designed for use with laboratory equipment. (Its language is actually referred to as G, but LabVIEW is so far the only implementation of G.) R began as a language for statistical computing, and is now widely used for general data analysis and beyond. It can be used interactively or built into an executable.

Whether it is better to use an interactive computing environment, a program written in a very high-level language, or write a program in a lower level language like C or Fortran depends on the task to be solved. Each has its domain of applicability. A single language or tool will be able to deal with a wide range of tasks, if needed, but will be inefficient or cumbersome for some of them. To be able to efficiently deal with a wide range of computational problems, it is advantageous to know several languages or tools from different parts of the spectrum: One high-level programming language for time-intensive number-crunching tasks (group I), one general-purpose computational environment for every-day calculations and data visualization (group II), and a few tools for data manipulation and processing (group III). This enables a scientist to choose a tool appropriate for the given task. Knowing multiple languages from the same category provides less of an advantage. On a practical level, use what you know, or use what the people you work with use. The endless number of special purpose software (group IV) can generally be treated as "no need to learn unless needed for a specific research project."

4.4 DATA VISUALIZATION

Graphics is an indispensable tool for data analysis, program validation, and scientific inquiry. We only want to avoid spending too much time on learning and coping with graphics software. Often, data analysis is exploratory. It is thus desirable to be able to produce a graph quickly and with ease. On the other hand, we also need to be able to produce publication quality figures. So we have to master *one* of those software well enough to make detailed adjustments. (One of those necessary adjustments, needed for figures in presentations, is to set the font size large enough so the audience can read the axes labels.)

Gnuplot is a simple and free graphics plotting program, www.gnuplot. info. It is quick to use and learn. A single command suffices to load *and* plot data, as will be demonstrated in an example below. *Origin* is one of many widely used proprietary software packages. *MatPlotLib* is a Python plotting library, `http://matplotlib.org`. All general-purpose math packages listed in Figure 4.1 include powerful graphing and visualization capabilities.

A simple demonstration of Gnuplot features:

```
# plot data in second column of myfile.dat versus first column
plot 'myfile.dat'

# the plotting range can be specified and errorbars drawn
plot [1:5][-1:1] 'myfile.dat' using 1:2:4 w errorbars

# perform arithmetic on the column entries and then plot
plot 'myfile.dat' using ($1*365.24):($3**2)

# logarithmic scale on the x-axis
set logscale x; replot
```

The pyplot module of MatPlotLib resembles Matlab commands, as in the following Python program:

```
import matplotlib.pyplot as plt
plt.plot([1,2,3,4],[2,7,8,2])
plt.xlabel('x (m)')
plt.show()
```

And `plt.xlabel('x (m)', fontsize='large')` increases the size of the axis label, so it is legible for those in the back row.

Recommended Reading: The legendary original reference book on C is Kernighan & Ritchie, *The C Programming Language*. Dennis Ritchie invented C. As an introductory book it is challenging, and it does not cover the updates to the C standards. A concise but complete coverage of Fortran is given by Metcalf, Reid, and Cohen, *Modern Fortran Explained*. There is also abundant instructional and reference material online. For Python, `www.python.org` includes extensive language documentation and pointers to further literature about scientific programming with Python with libraries such as NumPy, SciPy, MatPlotLib, and Pandas. Ample learning and reference material for NumPy and SciPy is also available at `www.scipy.org`.

EXERCISES

4.1 If you do not know any programming language yet, learn one. The following chapters will involve simple programming exercises.

4.2 *Gauss circle problem:* As a basic programming exercise, write a program that calculates the number of lattice points within a circle as a function of radius. In other words, how many pairs of integers (m, n) are there such that $m^2 + n^2 \leq r^2$. A finite resolution in r is acceptable. Submit the souce code of the program and a graph of the results.

4.3 With a programming language of your choice, read comma separated numbers from a file with an unknown number of rows. Submit the program source code.

4.4 Programming exercise with a language of your choice: Given a set of floating-point numbers between 0 and 10, sort them into bins of width 1, without explicitly looping through the bins for each number. Submit the program source code.

4.5 *Newton fractal:* As a programming exercise, a curious example of how complicated the domain of convergence for Newton's method can be is $z^3 - 1 = 0$ in the complex plane. The solutions to the equation are the cubic roots of $+1$. The domain of attraction for each of the three roots is a fractal. Write a program that produces this fractal, where $z = x + iy$ is the starting point of the iteration. If after 2000 steps the iteration has reached 1 or is close to 1, color the coordinate black, otherwise white. Hence, black will represent the set of initial conditions that converge to $+1$. Plot the results for $-1 \leq x \leq 1$, $-1 \leq y \leq 1$, at a resolution of 0.002.

## CHAPTER 5

Sample Problems; Building Conclusions

"Excellent computer simulations are done for a purpose."

Leo Kadanoff

"For it is a sad fact that most of us can more easily compute than think."

Forman Acton

Computing is a tool for scientific research. So we better get to the essence and start using it with a scientific goal in mind.

5.1 CHAOTIC STANDARD MAP

As an example, we study the following simple physical system. A freely rotating and frictionless rod is periodically kicked along a fixed direction, such that the effect of the kicking depends on the rod's position; see Figure 5.1. The equations for the angle α, measured from the vertical, and the angular velocity ω after each kick are

$$
\begin{aligned}
\alpha_{n+1} &= \alpha_n + \omega_n T \\
\omega_{n+1} &= \omega_n + K \sin \alpha_{n+1}
\end{aligned}
$$

The period of kicking is T and K is its strength. The iteration acts as a strobe effect for the time-continuous system, which records (α, ω) every time interval T, which is also the time interval between kicks.

For $K = 0$, without kicking, the rod will rotate with constant period; ω stays constant and α increases proportionally with time. For finite K, will the rod stably position itself along the direction of force ($\alpha = \pi$) or will it rotate

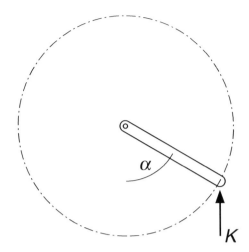

FIGURE 5.1 Freely rotating rod that is periodically kicked.

full turns forever? Will many weak kicks ultimately cause a big change in the behavior or not? Will it accumulate unlimited amounts of kinetic energy or not? These are nontrivial questions that can be answered with computations.

A program to iterate the above formula is only a few lines long. The angle can be restricted to the range 0 to 2π. If $\mathrm{mod}(\alpha, 2\pi)$ is applied to α at every iteration, it will cause a tiny bit of roundoff after each full turn (and none without a full turn), but it avoids loss of significant digits when α is large, as it will be after many full turns in the same direction. After a few thousand turns in the same direction, four significant digits are lost, because a few thousand times 2π is about 10^4. With 16 digits precision this is not all that bad.

To begin with, a few test cases are in order to validate the program, as short as it may be. For $K = 0$ the velocity should never change. A simple test run, $\omega = 0.2, 0.2, 0.2, ...$, confirms this. Without kicking $\alpha_{n+1} = \alpha_n + 2\pi T/T_r$, where $T_r = 2\pi/\omega$ is the rotation period of the rod. Between snapshots separated by time T, the angle α changes either periodically or, when T_r/T is not a ratio of integers, the motion is "quasi-periodic," because the rod never returns exactly to the same position.

A second test case is the initial values $\alpha_0 = 0$ and $\omega_0 = 2\pi/T$, that is, the rod rotates initially exactly at the kicking frequency and α should return to the same position after every kick, no matter how hard the kicking: $\alpha = 0, 0, 0, ...$. The program reproduces the correct behavior in cases we understand, so we can proceed to more complex situations.

For $K = 0.2$ the angular velocity changes periodically or quasi-periodically, as a simple plot of ω_n versus n, as in Figure 5.2, shows. Depending on the initial value, ω will remain of the same sign or periodically change its sign.

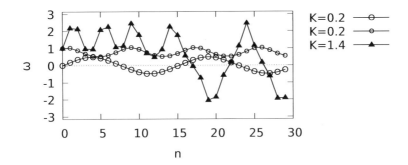

FIGURE 5.2 Angular velocity ω for the standard map for kicking strengths $K = 0.2$ and $K = 1.4$.

For stronger kicking, $K = 1.4$, the motion can be chaotic, for example $\alpha \approx$ 0, 1, 3.18, 5.31, 6.27, 0.94, 3.01, 5.27, 0.05, ... An example of the ω behavior is included in Figure 5.2. The rod goes around a few times, then reverses direction, and continues to move in an erratic fashion.

Plots of ω versus α reveal more systematic information about the behavior of the solutions. Figure 5.3 shows the phase space plots for several initial values for ω and α. Since α is periodic, the (α, ω) phase space may be thought of as a cylinder rather than a plane. Again we first consider the trivial case without kicking (Figure 5.3a). For $K = 0$, if ω is irrational it will ultimately sweep out all values of α, which appears as a horizontal straight line in the plot; if ω is a rational number there will only be a finite number of α values.

For small K, there is a small perturbation of the unforced behavior (Figure 5.3b). When the rod starts out slow and near the stable position $\alpha = \pi$, it perpetually remains near the stable position, while for high initial speeds the rod takes full turns forever at (quasi)periodically changing velocities. These numerical results suggest, but do not prove, that the motion, based on the few initial values plotted here, remains perpetually regular, because they lie on an invariant curve. The weak but perpetual kicking changes the behavior qualitatively, because now there are solutions that no longer go around, but bounce back and forth instead. Nevertheless, judged by Figure 5.3(b) all motions are still regular. Solutions that perpetually lie on a curve in phase space (or, in higher dimensions, a surface) are called "integrable." The theory discussion below will elaborate on integrable versus chaotic solutions.

The behavior for stronger K is shown in panel (c). For some initial conditions the rod bounces back and forth; others make it go around without ever changing direction. The two regions are separated by a layer where the motion is chaotic. The rod can go around several times in the same direction, and then reverse its sense of rotation. In this part of the phase space, the motion of the rod is perpetually irregular. The phase plot also reveals that this chaotic motion is bounded in energy, because ω never goes beyond a maximum value.

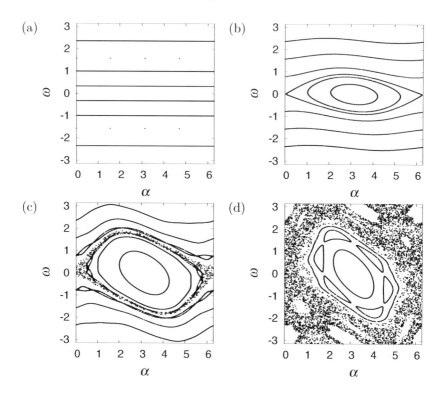

FIGURE 5.3 Angular velocity ω versus angle α for the standard map for different kicking strengths, (a) $K = 0$, (b) $K = 0.2$, (c) $K = 0.8$, and (d) $K = 1.4$. Nine initial conditions are used in each plot. Lines that appear to be continuous consist of many discrete points.

For strong kicking there is a "sea" of chaos (d). Although the motion is erratic for many initial conditions, the chaotic motion does not cover all possible combinations of velocities and angles. We see intricate structures in this plot. There are regular/quasi-periodic (integrable) solutions bound in α and ω. Then there are islands of regular behavior embedded in the sea of chaotic behavior. Simple equations can have complicated solutions, and although complicated they still have structure to them. A qualitative difference between (c) and (d) is that in (d) the chaotic motions are no longer bounded in energy. For some intermediate kicking strengths, this division in phase space disappears.

By the way, for strong kicking, the rod can accumulate energy without limit. In the figure, ω is renormalized to keep it in the range $-\pi/T$ to π/T. Physicist Enrico Fermi used a similar model to show that electrically charged particles can acquire large amounts of energy from an oscillating driving field. Unbounded growth of kinetic energy due to periodic forcing is thus also known as "Fermi acceleration". (Exercise 5.1 elaborates on this.)

Theory discussion: almost integrable systems*

The kicked rotator is a mechanical system without friction, so to those who know theoretical mechanics, it is a Hamiltonian system, and the pair of canonical variables (α, ω) forms a phase space. The map is known as the (Chirikov) "standard map" of chaotic Hamiltonian dynamics. It has the smallest number of variables possible for a Hamiltonian system (two) and simple functional dependencies. Although it may be a bit of a toy system, lessons drawn from studying its phase space structure apply to an extraordinarily wide range of Hamiltonian systems.

Kolmogorov-Arnold-Moser (KAM) theory is concerned with the stability of motions in Hamiltonian systems under small perturbations. (Not a perturbation to the initial conditions, but a small change to the governing equations, for example in the form of a small external force, such as the kicking of a rotator described above. Another example would be an asteroid that moves around our sun, but also feels the gravitational influence of the massive planet Jupiter, which is a small but perpetual influence. Does a feeble force exerted over a long time ultimately cause a major change in the trajectory or not?) KAM theory answers whether or not a small perturbation of a conservative dynamical system results in a lasting regular motion, and it gives conditions under which chaos is restricted. Qualitatively, the theorem is the following:

> KAM theorem: For a small perturbation of an integrable Hamiltonian system, invariant surfaces in phase space *continue to exist* for *most* initial conditions.

This is seen to be the case in Figure 5.3(b), where motions are regular; only the theorem allows for a region of chaotic behavior as in c). From Figure 5.3(a) to (b) the qualitative behavior of some solutions changes, since there are back-and-forth motions not present for the undriven case. But both types of solutions are still integrable, as predicted by the KAM theorem. A layer of chaotic behavior may be present in part (b), but too thin in phase space to show up in the plot. In Figure 5.3(c), the fraction of non-integrable solutions has grown to a layer of clearly discernible thickness. Some of the invariant curves are deformed and survive, while others are destroyed. KAM curves are deformed but remain impenetrable. Beyond a certain perturbation strength, this barrier breaks down (Figure 5.3(d)). This behavior is beyond the small perturbations described by the KAM theorem.

5.2 GRAVITATIONAL 3-BODY PROBLEM

The motion of two bodies due to their mutual gravitational attraction leads to orbits with the shape of circles, ellipses, parabolas, or hyperbolas. For three bodies an analytical solution is no longer possible. Nothing keeps us from exploring the gravitational interaction between three bodies numerically. For example, we may be interested in interstellar objects, arriving from far away, that will be captured into an orbit around the sun due to interaction with Jupiter, the heaviest planet in our solar system. The goal here is to get a perspective of the overall process of scientific computing: setting up the problem, validating the numerical calculations, and arriving at a reliable conclusion. The problem will only be outlined, so we can move from step to step more quickly.

We begin by writing down the governing equations. The acceleration of the bodies due to their gravitational interaction is given by

$$\frac{d^2\mathbf{r}_i}{dt^2} = -G\sum_{j\neq i} m_j \frac{\mathbf{r}_i - \mathbf{r}_j}{|\mathbf{r}_i - \mathbf{r}_j|^3}$$

where t is time, G is the gravitational constant, the sum is over all bodies, and m and $\mathbf{r}(t)$ are their masses and positions. We could use polar or cartesian coordinates. With three bodies, polar coordinates do not provide the benefit it would have with two bodies, so cartesian is a fine choice.

The next step is to simplify the equations. The motion of the center of mass can be subtracted from the initial conditions, so that all bodies revolve relative to the center of mass at a reduced mass. This reduces the number of equations, in two dimensions, from 6 to 4. In the following we will consider the "restricted three-body problem": The first mass (the Sun) is much heavier than the other two, and the second mass (Jupiter) is still much heavier than the third, so the second mass moves under the influence of the first, and the third mass moves under the influence of the other two. Since the first body is so heavy that it does not accelerate, it makes subtracting the center of mass superfluous, and we are left with equations of motion for the second and third body only.

The numerical task is to integrate the above system of ordinary differential equations (ODEs) that describe the positions and velocities of the bodies as a function of time. What capabilities does a numerical ODE solver need to have for this task? As the object arrives from far away and may pass close to the sun, the velocities can vary tremendously over time, so an adaptive time step is a huge advantage; otherwise, the time integration would have to use the smallest timestep throughout.

The problem of a zero denominator, $r_i = r_j$, arises only if the bodies fall straight into each other from rest or if the initial conditions are specially designed to lead to collisions; otherwise the centrifugal effect will avoid this problem for pointlike objects.

The masses and radii in units of kilograms and meters are large numbers; hence, we might want to worry about overflow in intermediate results. For example, the mass of the sun is 2×10^{30}kg, G is 6.67×10^{-11}m^3/kg·s, and the body starts from many times the Earth-sun distance of 1.5×10^{11}m. We cannot be sure of the sequence in which the operations are carried out. If G/r^3, the worst possible combination (Exercise 3.2), is formed first, the minimum exponent easily becomes as low as -44. The exponent of a single-precision variable (-38) would underflow. Double-precision representation is safe (-308).

In conclusion, this problem demands an ODE solver with variable step size and double-precision variables. Next we look for a suitable implementation. A general-purpose computational environment is easily capable of solving a system of ODEs of this kind, be it Python, Octave, Matlab, IDL, Mathematica, or another. (The name of the command may be ode, odeint, lsode, ode45, rk4, NDSolve, or whatnot). Or, in combination with a lower-level language, pre-written routines for ODE integration can be used. Appendix B provides a list of available repositories where implementations to common numerical methods can be found. We enter the equations to be solved, along with initial coordinates and initial velocity, and compute the trajectories. Or rather, we find an ODE integrator, learn how to use it, then enter the equations to be solved, and so forth.

The next step is clear: validation, validation, validation. As a first test, for the two-body problem, we can choose the initial velocity for a circular orbit. The resulting trajectory is plotted in Figure 5.4(a) and is indeed circular. In fact, in this plot the trajectory goes around in a circular orbit one hundred times, but the accuracy of the calculation and the plotting is so high that it appears to be a single circle.

Figure 5.4(b) shows the total energy (potential energy plus kinetic energy), which must be constant with time. This plot is only good enough for the investigator's purpose, because it has no axes labels, so let it be said that it is the total energy as a function of time. For the initial values chosen, the energy is -0.5. At this point it has to be admitted that this model calculation was carried out with $G = 1$, $m_{\text{Sun}} = 1$, $m_{\text{Jupiter}} = 1/1000$, and $r_{\text{Jupiter}} = 1$. Variations in energy occur in the sixth digit after the decimal point. Even after hundreds of orbits the error in the energy is still only on the order of 10^{-6}. Any numerical solution of an ODE will involve discretization errors, so this amounts to a successful test of the numerical integrator. What could be a matter of concern is the systematic trend in energy, which turns out to continue to decrease even beyond the time interval shown in the plot. The numerical orbit spirals slowly inward, in violation of energy conservation. The Earth has moved around the Sun many times,–judged by its age, over four and a half billion times,–so if we were interested in such long integration periods, we would face a problem with the accuracy of the numerical solver.

Another case that is easy to check is a parabolic orbit. We can check that the resulting trajectory "looks like" a parabola, but without much extra work the test can be made more rigorous. A parabola can be distinguished from a hyperbolic or strongly elliptic trajectory by checking that its speed approaches zero as the body travels toward infinity.

Trajectories in a $1/r$ potential repeat themselves. This is not the case for trajectories with potentials slightly different from $1/r$. This is how Isaac Newton knew that the gravitational potential decays as $1/r$, without ever measuring the gravitational potential as a function of r. The mere fact that the numerical solution is periodic, as in an elliptic orbit, is a nontrivial validation.

Once we convinced ourselves of the validity of the numerical solutions to

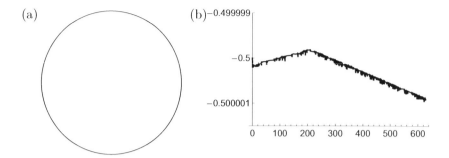

FIGURE 5.4 Numerical solutions for the gravitational interaction of two bodies. (a) A circular orbit for testing purposes. The orbit has actually undergone 100 revolutions, demonstrating the accuracy of the numerical solution. (b) Total energy as a function of time, where the orbital period is 2π. The error is acceptably small.

the two-body problem, we can move on to three bodies. We will consider the restricted three-body problem in a plane. The mass of Jupiter is about 1,000 times lower than that of the Sun, and the mass of the third body is negligible compared to Jupiter's.

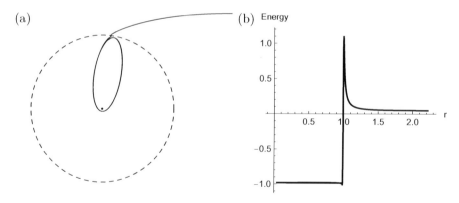

FIGURE 5.5 Numerical solutions for the gravitational interaction of three bodies. (a) Motion of three bodies (center point, solid line, dashed line) with very different masses. A body from far away is captured into a periodic orbit. (b) Total energy of the third body as a function of distance from center.

After all these steps, we are prepared to address the original issue of capture of a visiting interstellar object. Figure 5.5(a) shows an example of three-body motion, where one body from far away comes in and is deflected by the orbiting planet such that it begins to orbit the sun. Its initial conditions are such that without the influence of the planet the object would escape to infinity. Jupiter captures comets in this manner.

For deeper insight, the total energy as a function of distance from the sun is shown in Figure 5.5(b). The horizontal axis is distance from the sun, and the third body starts from a large distance, so as time proceeds the graph has to be followed from right to left. This plot shows that the initial total energy is positive; in other words the body's initial kinetic energy is larger than its initial potential energy, so without the planet, the body would ultimately escape the sun's gravity. The third body loses energy at the same distance as the second body's orbit and transitions into a gravitationally bound orbit. During this interaction the variable step size of the numerical integrator is again of great benefit. Ultimately, the body ends up with negative total energy, which implies it is gravitationally bound to the sun.

Is this numerical answer reliable? After all, there is rapid change during the close encounter of Jupiter with the third body. To check, we can conduct a number of sensitivity tests. For a sensitivity test, we do not have to carry out a bigger calculation, in case that strains the computer, but we can carry out a smaller calculation. A respectable ODE integrator will have some sort of accuracy parameter or goal the user can specify. We can decrease this parameter to see whether the answer changes. For our standard parameters the energy is -0.9116. To test the result's sensitivity to the accuracy goal of the integrator, we reduce this parameter by an order of magnitude, and obtain -0.9095. The result has changed. Now it is appropriate to increase the accuracy goal by an order of magnitude and obtain -0.9119. There is still a change compared to -0.9116, but it is smaller; it takes place in the fourth digit after the decimal point instead of the second digit. It is okay for the answer to change with the accuracy of the calculation, as long as it approaches a fixed value, especially since the main conclusion, that an object with these specific initial conditions gets captured, only requires that the total energy be below zero. A "convergence test", where we check that the answer approaches a constant value, is a more elaborate version of a sensitivity test. Chapter 6 will treat convergence tests in detail.

The single most important point to remember from this exercise is that we proceed from simple to complex and validate our numerical computations along the way. The pace and detail at which this should proceed depends on the familiarity and confidence of the user in the various components of the task. Computational scientists not only write programs, they also build conclusions. In this aspect the expectation from computation-based research is no different from experiment-based research, which demands control experiments, or theory-based research, where equations must be consistent with known answers in special cases.

EXERCISES

5.1 For the kicked rotator $(\alpha_{n+1} = \alpha_n + \omega_n T, \; \omega_{n+1} = \omega_n + K \sin \alpha_{n+1})$, determine how fast the energy $E = \omega^2/2$ grows with time n. Consider the ensemble average and large kicking strength ($K \geq 4$). You need to consider only one value of K; the answer is supposedly independent of K beyond this value. Take initial values that lead to chaotic motions, that is, find a way to exclude integrable solutions from the average. The function you fit should have a physically reasonable asymptotic behavior.

Submit: 1) a plot of the ensemble-averaged energy as a function of time and a functional fit to this graph, 2) a description of how you avoided the integrable solutions, and 3) basic info such as the value of K used, how many initial values were chosen, and how they were chosen.

5.2 Solve Kepler's equation with Newton's method. The Kepler equation is $M = E - e \sin E$, where the so-called "mean anomaly" M is linear in time, E is the eccentric anomaly, and e is the eccentricity of the orbit. The distance from the sun is $r = a(1 - e \cos E)$, so solving Kepler's equation provides r for a given time.

a. Write a program that solves Kepler's equation for any M. Use a reasonable criterion to decide how many iterations are necessary.

b. Test the program with exact solutions. Use $e = 0.9671$, appropriate for Halley's comet.

c. Calculate the time average of $(a/r)^2$ to at least three significant digits. The mean solar flux is proportional to this quantity. (This average can be obtained analytically, but the task here is to obtain it numerically.)

## CHAPTER 6

Approximation Theory

While mathematical functions can be defined on a continuous variable, any numerical representation is limited to a finite number of values. This discretization of the continuum is the source of profound issues for numerical interpolation, differentiation, and integration.

6.1 DIFFERENTIATION: FINITE DIFFERENCES

The derivative of a function can be approximated in terms of the difference between two nearby values. Here we consider this method of approximation systematically. A function is locally described by its Taylor expansion:

$$f(x+h) = f(x) + f'(x)h + f''(x)\frac{h^2}{2} + \ldots + f^{(n)}(x)\frac{h^n}{n!} + f^{(n+1)}(x+\vartheta)\frac{h^{n+1}}{(n+1)!}$$

The very last term is evaluated at $x + \vartheta$, which lies somewhere between x and $x + h$. Since ϑ is unknown, this last term provides a bound on the error when the series is truncated after n terms. For example, $n = 0$ implies that $|f(x+h) - f(x)| \leq Mh$, where $M = \max_{0 \leq \vartheta \leq h} |f'(x+\vartheta)|$. And $n = 1$ implies that $f(x+h) - f(x) = f'(x)h + f''(x+\vartheta)h^2/2$ or

$$f'(x) = \frac{f(x+h) - f(x)}{h} + O(h).$$

A function is said to be "of order p", $O(h^p)$, when for sufficiently small h its absolute value is smaller than a constant times h^p. In our example, that is the case if the constant is $(1/2)\max_{0 \leq \vartheta \leq h} |f''(x+\vartheta)|$.

The derivative of a function can be approximated by a difference over a finite distance, $f'(x) \approx [f(x+h) - f(x)]/h$, the "forward difference" formula, or $f'(x) \approx [f(x) - f(x-h)]/h$, the "backward difference" formula. The Taylor expansion can be used to verify finite difference expressions for derivatives and to obtain an error bound: $f'(x) = [f(x+h) - f(x)]/h + O(h)$ and

$$f'(x) = \frac{f(x) - f(x-h)}{h} + O(h).$$

Another possibility is the "center difference"

$$f'(x) = \frac{f(x+h) - f(x-h)}{2h} + O(h^2).$$

The center difference is accurate to $O(h^2)$, not just $O(h)$ as the one-sided differences are, because the f'' terms in the Taylor expansions of $f(x+h)$ and $f(x-h)$ cancel:

$$
\begin{aligned}
f(x+h) &= f(x) + f'(x)h + f''(x)\frac{h^2}{2} + f'''(x+\vartheta_+)\frac{h^3}{3!} \\
f(x-h) &= f(x) - f'(x)h + f''(x)\frac{h^2}{2} - f'''(x+\vartheta_-)\frac{h^3}{3!}
\end{aligned}
$$

The Taylor expansion is written to one more order than may have been deemed necessary; otherwise, the cancellation of the second-order time would have gone unnoticed.

The center point, $f(x)$, is absent from the difference formula, and at first sight this may appear awkward. A parabola fitted through the three points $f(x+h), f(x)$, and $f(x-h)$ undoubtedly requires $f(x)$. However, it is easily shown that the slope at the center of such a parabola is independent of $f(x)$ (Exercise 6.1). Thus, it makes sense that the center point does not appear in the finite difference formula for the first derivative.

The second derivative can also be approximated with a finite difference formula, $f''(x) \approx c_1 f(x+h) + c_2 f(x) + c_3 f(x-h)$, where the coefficients c_1, c_2, and c_3 can be determined with Taylor expansions. This is a general method to derive finite difference formulae. Each order of the Taylor expansion yields one equation that relates the coefficients to each other. After some calculation, we find

$$f''(x) = \frac{f(x-h) - 2f(x) + f(x+h)}{h^2} + O(h^2).$$

With three coefficients, c_1, c_2, and c_3, we only expect to match the first three terms in the Taylor expansions, but the next order, involving $f'''(x)$, vanishes automatically. Hence, the leading error term is $O(h^4)/h^2 = O(h^2)$. A mnemonic for this expression is the difference between one-sided first derivatives:

$$f''(x) \approx \frac{\frac{f(x+h)-f(x)}{h} - \frac{f(x)-f(x-h)}{h}}{h}$$

It is called a mnemonic here, because it does not reproduce the order of the error term, which would be $O(h)/h = O(1)$, although it really is $O(h^2)$.

With more points (a larger "stencil") the accuracy of a finite-difference approximation can be increased, at least as long as the high-order derivative that enters the error bound is not outrageously large. The error term involves a higher power of h, but also a higher derivative of f. Reference books provide the coefficients for various finite-difference approximations to derivatives on one-sided and centered stencils of various widths.

6.2   VERIFYING THE CONVERGENCE OF A METHOD

Consider numerical differentiation with a simple finite difference:

$$u(x) = \frac{f(x+h) - f(x-h)}{2h}$$

With a Taylor expansion we can immediately verify that $u(x) = f'(x) + O(h^2)$. For small h, this formula provides therefore an approximation to the first derivative of f. When the resolution is doubled, the discretization error, $O(h^2)$, decreases by a factor of 4. Since the error decreases with the square of the interval h, the method is said to converge with "second order." In general, when the discretization error is $O(h^p)$ then p is called the "order of convergence" of the method.

The resolution h can be expressed in terms of the number of grid points N, which is simply inversely proportional to h. To verify the convergence of a numerical approximation, the error can be defined as some overall difference between the solution at resolution $2N$ and at resolution N that we denote with u_{2N} and u_N. Ideal would be the difference to the exact solution, but the solution at infinite resolution is usually unavailable, because otherwise we would not need numerics. "Norms" (denoted by $\|\cdot\|$) provide a general notion of the magnitude of numbers, vectors, matrices, or functions. One example of a norm is the root-mean-square

$$\|y\| = \sqrt{\frac{1}{N} \sum_{j=1}^{N} (y(jh))^2}$$

Norms of differences therefore describe the overall difference, deviation, or error. The ratio of errors, $\|u_N - u_{N/2}\| / \|u_{2N} - u_N\|$, must converge to 2^p, where p is the order of convergence.

$$\lim_{N \to \infty} \frac{\|u_N - u_{N/2}\|}{\|u_{2N} - u_N\|} \to 2^p$$

Table 6.1 shows a convergence test for the center difference formula shown above applied to an example function. The error $E(N) = \|u_{2N} - u_N\|$ becomes indeed smaller and smaller with a ratio closer and closer to 4.

The table is all that is needed to verify convergence. For deeper insight, however, the errors are plotted for a wider range of resolutions in Figure 6.1. The line shown has slope -2 on a log-log plot and the convergence is overwhelming. The bend at the bottom is the roundoff limitation. Beyond this resolution the leading error is not discretization but roundoff. If the resolution is increased further, the result becomes less accurate. For a method with high-order convergence this roundoff limitation may be reached already at modest resolution. A calculation at low resolution can hence be more accurate than a calculation at high resolution!

TABLE 6.1 Convergence test for the first derivative of the function $f(x) = \sin(2x - 0.17) + 0.3\cos(3.4x + 0.1)$ in the interval 0 to 1 and for $h = 1/N$. The error (second column) decreases with increasing resolution and the method therefore converges. Doubling the resolution reduces the error by a factor of four (third column), indicating the finite-difference expression is accurate to second order.

N	$E(N)$	$E(N/2)/E(N)$
20	0.005289	
40	0.001292	4.09412
80	0.0003201	4.03556
160	7.978E-05	4.01257

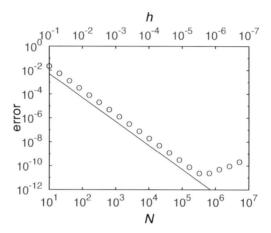

FIGURE 6.1 Discretization and roundoff errors for a finite-difference formula. The total error (circles), consisting of discretization and roundoff errors, decreases with resolution N until roundoff errors start to dominate. For comparison, the solid line shows the theoretical discretization error, proportional to $1/N^2$ or h^2, with an arbitrary prefactor.

To understand the plot more thoroughly, we check for quantitative agreement. In the convergence test shown in the figure, double-precision numbers are used with an accuracy of about 10^{-16}. The function values used for Figure 6.1 are around 1, in order of magnitude, so that absolute and relative errors are approximately the same. The roundoff limitation occurs in this example at an accuracy of 10^{-11}. Why? In the formation of the difference $f(x + h) - f(x - h)$ a roundoff error of about 10^{-16} is introduced, but to obtain u, it is necessary to divide by $2h$, enhancing the absolute error because h is a small number. In the figure the maximum accuracy is indeed approximately $10^{-16}/(2 \times 5 \times 10^{-6}) = 10^{-11}$. The total error is the sum of discretization error and roundoff error $O(h^2) + O(\varepsilon/h)$, where $\varepsilon \approx 10^{-16}$. If we are sloppy and consider the big-O not merely as a smaller than a constant times its argument, but as an asymptotic limit, then the expression can be differentiated: $O(2h) - O(\epsilon/h^2)$. The total error is a minimum when $h = O(\varepsilon^{1/3}) = O(5 \times 10^{-6})$. This agrees perfectly with what is seen in Figure 6.1.

The same convergence test can be used not only for derivatives, but also for integration schemes, differential equations, and anything else that ought to become more accurate as a parameter, such as resolution, is changed.

6.3 NUMERICAL INTEGRATION: ILLUSIONS ABOUT WHAT LIES BETWEEN

The simplest way of numerical integration is to sum up function values. Rationale can be lent to this procedure by thinking of the function values $f_j = f(x_j)$ as connected with straight lines. Let f_j denote the function at $x_j = x_0 + jh$. The area of the first trapezoidal segment is, using simple geometry, $(f_0+f_1)/2$. The area under the piecewise linear graph from x_0 to x_N is

$$\int_{x_0}^{x_N} f(x)dx \approx \frac{f_0 + f_1}{2}h + \frac{f_1 + f_2}{2}h + \ldots$$

$$= \left(\frac{f_0}{2} + f_1 + \ldots + f_{N-1} + \frac{f_N}{2} \right) h$$

which is indeed the sum of the function values. The boundary points carry only half the weight. This summation formula is called the "composite trapezoidal rule."

Instead of straight lines it is also possible to *imagine* the function values are interpolated with quadratic polynomials. Fitting a parabola through three points and integrating, one obtains

$$\int_{x_0}^{x_2} f(x)dx \approx \frac{h}{3} (f_0 + 4f_1 + f_2)$$

For a parabola the approximate sign becomes an exact equality. This integration formula is well-known as "Simpson's rule." Repeated application of Simpson's rule leads to

$$\int_{x_0}^{x_N} f(x)dx \approx \frac{h}{3} [f_0 + 4f_1 + 2f_2 + 4f_3 + 2f_4 + \ldots + 4f_{N-1} + f_N].$$

An awkward feature of this "composite Simpson formula" is that function values are weighted unequally, although the grid points are equally spaced.

There is an exact relation between the integral and the sum of a function, known as "Euler-Maclaurin summation formula":

$$\int_a^b f(x)dx = h\sum_{j=1}^{N-1} f(a+jh) + \frac{h}{2}\left(f(a)+f(b)\right) +$$

$$-\sum_{j=1}^{m} h^{2j}\frac{B_{2j}}{(2j)!}\left(f^{(2j-1)}(b) - f^{(2j-1)}(a)\right) +$$

$$-h^{2m+2}\frac{B_{2m+2}}{(2m+2)!}(b-a)f^{(2m+2)}(\vartheta),$$

where B_k are the Bernoulli numbers, $h = (b-a)/N$, and ϑ lies somewhere between a and b. The Bernoulli numbers are mathematical constants; the first few of them are $B_2 = 1/6$, $B_4 = -1/30$, $B_6 = 1/42$, ... Bernoulli numbers with odd indices are zero.

The Euler-Maclaurin summation formula can be used to determine the error when an integral is approximated by a sum, just as the Taylor expansion provided the error of a finite-difference formula. For $m = 0$,

$$\int_a^b f(x)dx = h\left(\frac{f_0}{2} + f_1 + \ldots + f_{N-1} + \frac{f_N}{2}\right) - h^2\frac{B_2}{2!}(b-a)f''(\vartheta).$$

The first order of the Euler-Maclaurin summation formula *is* the trapezoidal rule and the error for trapezoidal integration is $-h^2(b-a)f''(\vartheta)/12$. Therefore, the error is $O(h^2)$.

The Euler-Maclaurin summation formula for $m = 1$ is

$$\int_a^b f(x)dx = h\left(\frac{f_0}{2} + f_1 + \ldots + f_{N-1} + \frac{f_N}{2}\right) +$$

$$-h^2\frac{B_2}{2!}(f'(b) - f'(a)) - h^4\frac{B_4}{4!}(b-a)f^{(4)}(\vartheta).$$

It is now apparent that the leading error in the composite trapezoidal rule, $O(h^2)$, arises from the boundaries only, not from the interior of the domain. If $f'(a)$ and $f'(b)$ are known or if they cancel each other, the integration error is only $-h^4(b-a)f^{(4)}(\vartheta)/720$ or $O(h^4)$. Incidentally, the right-hand side has terms of size $O(h)$, $O(h^2)$, and $O(h^4)$, but there is no $O(h^3)$ term.

The composite Simpson formula can be derived by using the Euler-Maclaurin summation formula with spacings h and $2h$. The integration error obtained in this way is $h^4(b-a)\left[\frac{1}{3}f^{(4)}(\vartheta_1) - \frac{4}{3}f^{(4)}(\vartheta_2)\right]/180$. Since the error is proportional to $f^{(4)}$, applying Simpson's rule to a *cubic* polynomial yields the integral exactly, although it is derived by integrating a *quadratic* polynomial. That is the consequence of the lack of an $O(h^3)$ term.

The 4th order error bound in the Simpson formula is larger than in the trapezoidal formula. The Simpson formula is only more accurate than the trapezoidal rule, because it better approximates the boundary regions. Away from the boundaries, the Simpson formula, the method of higher order, yields less accurate results than the trapezoidal rule, which is the penalty for the unequal coefficients. At the end, simple summation of function values is an excellent way of integration in the interior of the domain. Thinking that parabolas better approximate the area under the graph than straight lines is an illusion.

6.4 NOTIONS OF ERROR AND CONVERGENCE*

"Norms" (denoted by $\|\cdot\|$) provide a general notion of the magnitude of numbers, vectors, matrices, or functions. The norms of differences are a measure of deviation or error. The norm of a single number is simply its absolute value. For vectors there are various possibilities. Here are a few:

$$\|y\|_1 = \frac{1}{N}\sum_{j=1}^{N}|y_j| \quad \text{1-norm}$$

$$\|y\|_2 = \sqrt{\frac{1}{N}\sum_{j=1}^{N}y_j^2} \quad \text{2-norm, root-mean-square}$$

$$\|y\|_\infty = \max_{j=1\ldots N}|y_j| \quad \text{max-norm}$$

Often these norms are defined without the $1/N$ factors. Either way, they are norms.

The following properties define a norm:

1. $\|cv\| = |c|\|v\|$ homogeneity

2. $\|u + v\| \le \|u\| + \|v\|$ triangular inequality

3. if $\|v\| = 0$, then $v = 0$

The first defining property is proportionality, but with an absolute value. It also implies positiveness, $\|v\| \ge 0$ always. The second defining property captures a notion of distance expected in all dimensions. The first property already implies that $\|0\| = 0$, whereas the last property requires that the reverse also be true.

Certainly, when v is a single number, the absolute value is a norm, because $|cv| = |c| \cdot |v|$, $|u + v| \le |u| + |v|$, and $|v| = 0$ implies $v = 0$. In fact, norms can be considered a generalization of the "absolute value" to other mathematical objects. That is why it makes sense to denote them with a double bar $\|.\|$ rather than a single bar $|.|$. It would be easy to demonstrate that each of the examples of vector norms above satisfies these three properties.

A sequence v_n is said to converge to zero if $\|v_n\| \to 0$ as $n \to \infty$. Now here is a helpful property about the norms of vectors: if a sequence converges in one norm, it also converges in all other norms. Certainly, the 1-norm is only zero when each element is zero, and the same goes for the 2-norm, and the max-norm as defined above (the number of elements N is fixed).

For functions, analogous definitions can be used:

$$\|y\|_1 = \int |y| dx \quad \text{1-norm}$$

$$\|y\|_2 = \left(\int y^2 dx\right)^{\frac{1}{2}} \quad \text{2-norm}$$

$$\|y\|_\infty = \sup_x |y| \quad \text{supremum norm}$$

The 1-norm is the area under the graph. The 2-norm is the integral over the square of the deviations. The supremum norm is sometimes also called ∞-norm, maximum norm, or uniform norm. (Those who do not remember the definition of supremum as the least upper bound can think of it as the maximum.) Ultimately, these function values will need to be represented by discretely sampled values, but the difference to the vector norms above is that the number of elements is not fixed; it will increase with the resolution at which the function is sampled.

For the norm of functions it is no longer true that convergence in one norm implies convergence in all norms. If y unequals zero for only one argument, the area under the function is zero, $\|y\|_1 = 0$, but the supremum (maximum) is not, $\|y\|_\infty > 0$. Hence, for functions there is more than one notion of "approximation error".

Norms, which reduce a vector of deviations to a single number that characterizes the deviation, can also be used to define a condition number, which is simply the proportionality factor between the norm of the input error and the norm of the output error.

Norms can also be calculated for matrices, and there are many types of norms for matrices, that is, functions with the three defining properties above. All that will be pointed out here is that matrix norms can be used to calculate the condition number of a linear system of equations.

Brainteaser: It is interesting to compare numerical convergence tests with the formal mathematical notion of convergence. Our convergence test in section 6.2 showed that $\|u_{2N} - u_N\| \to 0$ as the resolution N goes to infinity (roundoff ignored). Does this mean $\lim_{N\to\infty} \|u_N - u\| \to 0$, where u is the exact, correct answer? (Hint: Cauchy sequences)

6.5 POLYNOMIAL INTERPOLATION

Interpolation is concerned with how to approximate a function between two or more known values. Approximation by a polynomial springs to mind; after all, the value of a polynomial can easily be calculated for any x in a finite number of steps. Figure 6.2 displays a fit with a polynomial which passes exactly through equally spaced points on the abscissa. Although the polynomial goes exactly through every point as demanded, it badly represents the function in *between* grid points. The center part of the function is well approximated by the polynomial, but the polynomial oscillates excessively near the boundary of the domain. This oscillatory problem is known as the "Runge phenomenon." Consequently, using a polynomial of high degree to approximate a function can be a bad idea.

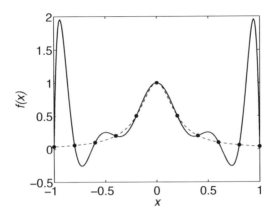

FIGURE 6.2 The Runge phenomenon. A polynomial (solid line) is fitted through 11 equally spaced points of the function $f(x) = 1/(1 + 25x^2)$ (dash line). Close to the boundaries of the domain, there are large deviations between the polynomial and the function.

When a function should be approximated over the entire domain, it is better to use many polynomials of low degree than a single polynomial of high degree. Piecewise polynomial fits, where the polynomials are joined smoothly with each other, are called "splines." For example, cubic polynomials which are joined such that the first derivative matches are a common choice, known as cubic spline interpolation. Interpolation by splines does not suffer from the oscillatory or stiffness problem that polynomials of high degree may display.

According to the Weierstrass Approximation Theorem, it is possible to approximate any continuous function on a closed interval with a single polynomial, such that the maximum difference between the function and its approximation is arbitrarily small. More formally, if f is a continuous function on $[a, b]$, then for any given $\epsilon > 0$, there exists a polynomial p such that $|f(x) - p(x)| < \epsilon$ for all x in $[a, b]$. This is not a contradiction to the above example, because best-approximation polynomials do not need to coincide with the function at equally spaced intervals. Best fitting polynomials intersect the function at points that are more densely spaced toward the interval boundary.

(We can ask the question: Which polynomial of degree n, or at most n, minimizes the deviation to a given function f on an interval $[a, b]$: $\|f - p_n\|_\infty$? In other words, we seek the minimum over all polynomial coefficients of the supremum norm—the minimum of a maxium. The polynomial of best approximation is in general difficult to find, but some types of polynomials, such as so-called Chebyshev polynomials, are very good at this. And the roots of Chebyshev polynomials are spaced more closely toward the two edges.)

An alternative to approximation with polynomials is to approximate a function $f(x)$ with a series of trigonometric functions, a topic that will be dealt with in chapter 7.

An overarching lesson from the examples in this chapter is that high-order approximations can be worse than low-order approximations. For differentiation, a higher order expansion means the order of the error term is higher, but because more numbers of about equal size are subtracted from each other, the limitation to accuracy from number truncation kicks in earlier. For integration, it is flatly wrong to think that a higher degree polynomial better approximates the area under a graph than piecewise straight lines would; only the errors caused by the two boundaries make it appear so. For interpolation, a high degree polynomial can widely oscillate. These are some of the ways we can be fooled.

On the other hand, the discretization error sometimes is smaller than we may have anticipated, due to cancellation of error terms, as for derivatives evaluated on symmetric stencils. Also, with the right type of polynomials, or piecewise polynomials, an arbitrarily accurate approximation to a continuous function can always be accomplished.

Recommended Reading: Chapter 25 of Abramowitz & Stegun, *Handbook of Mathematical Functions*, contains a great number of finite-difference expressions, http://people.math.sfu.ca/~cbm/aands/. Standard textbooks on numerical analysis include approximation theory. For a more advanced treatment, see Trefethen, *Approximation Theory and Approximation Practice*.

EXERCISES

6.1 Show mathematically that a parabola $p(x)$ that passes through three equally spaced points $y_1 = p(-h), y_2 = p(0), y_3 = p(h)$ has slope $(y_3 - y_1)/(2h)$ at the center point $x = 0$. In other words, the slope at y_2 does not depend on y_2.

6.2 Finite-difference expression for unequally spaced points.

a. Based on a stencil of three points $(0, x_1, x_2)$, find an approximation of the first derivative at $x = 0$, that is, find the coefficients for $f'(0) = c_0 f(0) + c_1 f(x_1) + c_2 f(x_2)$.

b. Determine the order of the error term.

c. Implement and verify with a numerical convergence test that the finite-difference approximation converges.

d. Also demonstrate that it converges at the anticipated rate.

6.3 Show that in the neighborhood of a simple root, Newton's method converges quadratically.

6.4 Error in numerical integration
   a. Implement a simple trapezoidal integrator.
   b. Numerically evaluate $\int_0^1 \cos(2\pi(1 - x^2) + \frac{1}{2})dx$, and determine the order of convergence as a function of spatial resolution.
   c. For $\int_0^1 \cos(2\pi(1-x^2))dx$, the integrand has vanishing first derivatives on both boundaries. Again determine the order of convergence.

6.5 Carry out Exercise 5.2 and add a convergence test for part c.

6.6 Derive a formula for the integral of a cubic polynomial in the interval 0 to b. Demonstrate that applying Simpson's rule to it will give the exact integral.

6.7 Show that for a vector y with n elements

$$\|y\|_\infty \leq \|y\|_2 \leq \|y\|_1 \leq \sqrt{n}\|y\|_2 \leq n\|y\|_\infty$$

6.8 Learn how to carry out spline interpolation within a software tool or language of your choice, and demonstrate that the function in Figure 6.2, $f(x) = 1/(1 + 25x^2)$ on the interval -1 to $+1$, can be approximated without the large oscillations.

## CHAPTER 7

Other Common Computational Methods

Here we describe methods and issues for some of the most commonplace computational problems that have not already been described. Root finding was discussed in chapter 2 and methods of linear algebra will be described in chapter 10. This chapter is dedicated to a handful of other problems so important that they cannot be omitted.

7.1 FITTING GRAPHS TO DATA IN THE AGE OF COMPUTATION

Fitting straight lines by the least-square method is straightforward: we minimize the quadratic deviations $E = \sum_i (y_i - a - bx_i)^2$, where x_i and y_i are the data and a and b are, respectively, the intercept and slope of a straight line, $y = a + bx$. This is the "Method of Least Squares". The extremum conditions $\partial E/\partial a = 0$ and $\partial E/\partial b = 0$ lead to the linear equations

$$\sum_i (a + bx_i - y_i) = 0 \quad \text{and} \quad \sum_i x_i(a + bx_i - y_i) = 0$$

which can be explicitly solved for a and b. The results are the well-known formulae

$$a = \frac{(\sum y_i)\sum x_i^2 - (\sum x_i)\sum x_i y_i}{N \sum x_i^2 - (\sum x_i)^2} \quad \text{and} \quad b = \frac{N \sum x_i y_i - (\sum x_i)(\sum y_i)}{N \sum x_i^2 - (\sum x_i)^2}$$

where the sums go from $i = 1, ..., N$.

The popularity of linear regression is partially due to do the computational convenience the fit parameters can be obtained with: there is an explicit formula for the coefficients. Minimizing with another weighting function would mean the equations for a and b are no longer linear. To convince ourselves of this, suppose the error or deviation d is weighted by an arbitrary function $w[d] \neq d^2$. Just once will we use square brackets to indicate the argument of a function to avoid confusion with multiplication. The error is $E = \sum_i w[a + bx_i - y_i]$, and $\partial E/\partial a = 0$ and $\partial E/\partial b = 0$ yield

$$\sum_i w'[a + bx_i - y_i] = 0 \quad \text{and} \quad \sum_i x_i w'[a + bx_i - y_i] = 0$$

For square deviations, $w = d^2$, the derivative $w' = 2d$, which reduces to the case above. But if w' is nonlinear, then two coupled nonlinear equations need to be solved to obtain the parameters a and b. As we learned in chapter 2, numerical nonlinear root-finding does not always guarantee a solution and there could be more than one local minimum for E. Alternatively, one can directly minimize the error E using numerical optimization (minima finding) methods, but conceptionally this approach has similar limitations as root-finding; the search could get stuck in a local minimum that is not the global minimum, so we cannot be certain the optimum found is global rather than only local.

This little exercise provides the following lessons: 1) Minimizing the square deviations from a straight line is a rare situation for which the fit parameters can be calculated with explicit formulae. Without electronic computers, it had to be that or eye-balling the best fit on a graph. 2) For regression with non-quadratic weights the fit obtained with a numerical search algorithm is not necessarily unique.

Minimizing the sum of the square deviations also has the following fundamental property: it yields the most likely fit for Gaussian distributed errors. We will prove this property, because it also exemplifies maximum likelihood methods. As a warm-up, we prove the following simpler statement: The sum of $(x_i - x)^2$ over a set of data points x_i is minimized when x is the sample mean. All we have to do is

$$\frac{\partial}{\partial x} \sum_i (x_i - x)^2 = 0$$

which immediately yields

$$\sum_i x_i = \sum_i x$$

and therefore x is the arithmetic mean.

Suppose the values are distributed around the model value with a probability distribution p. The sample value is y_i and the model value is a function $y(x_i; a, b, ...)$, where $a, b, ...$ are the fit parameters. A straight line fit would be $y(x_i; a, b) = a + bx_i$. We may abbreviate the deviation for data point i as $d_i = y(x_i; a, b, ...) - y_i$. The probability distribution is then $p(d_i)$, the same p for each data point. The maximum likelihood fit maximizes the product of all these probabilities $P = \Pi_i p_i$ where $p_i = p(d_i)$. An extremum with respect to a occurs when $\partial P/\partial a = 0$, and according to the product rule for differentiation

$$\frac{\partial P}{\partial a} = \frac{\partial}{\partial a} \Pi_i p_i = P \sum_i \frac{1}{p_i} \frac{\partial p_i}{\partial a}$$

Hence the extremum condition becomes

$$\sum_i \frac{\partial \ln p_i}{\partial a} = \frac{\partial}{\partial a} \sum_i \ln p_i = 0$$

In other words, $\Pi_i p_i$ is maximized when $\sum_i \ln p_i$ is maximized. When p is Gaussian, $p \propto \exp(-d^2/2\sigma^2)$ then $\sum_i \ln p_i = (\text{constant}) - 1/(2\sigma^2)\sum_i d_i^2$. Hence, the most likely parameters are those that minimize the sum of the square deviations, $\sum_i d_i^2$. The negative sign implies that if $\sum_i d_i^2$ is minimized, then $\sum_i \ln p_i$ and $\Pi_i p_i$ are maximized. This concludes the proof that minimizing the square of the deviations yields the most likely fit, if the errors are distributed Gaussian.

Some functions can be reduced to linear regression, e.g., $y^2 = \exp(x)$ implies that $\ln y$ depends linearly on x. If errors are distributed Gaussian then linear regression finds the most likely fit, but a transformation of variables spoils this property. So, although such a transformation readily allows to fit the data, this fit no longer minimizes the square deviations of the original variables.

Fits with quadratically weighted deviations are not particularly robust, since an outlying data point can affect it significantly. If the errors are indeed distributed Gaussian, then this is how it should be, but in reality the errors are often not Gaussian. Weighting proportional with distance, for instance, improves robustness.

7.2 FOURIER TRANSFORMS

Fourier transforms belong to the most used algorithms and have numerous applications. The Fourier Transform of signals describes the frequencies they are composed of. They are used to filter signals, by removing a specific frequency or a whole range of frequencies. There is also a fundamental relation between correlation functions and Fourier Transforms, further extending their application to realms of data analysis.

The Fourier Transform is defined by a continuous integral, $\hat{F}(k) = \int f(x)\exp(-ikx)dx$. In practice, we are limited to discretely sampled points and a finite number of elements. A discrete Fourier Transform with uniformly spaced points $j\Delta x$ and uniformly spaced frequencies $k\Delta k$ is defined by

$$\hat{F}_k = \sum_{j=0}^{N-1} f_j e^{-ikj\Delta x \Delta k}$$

If the slowest/longest period ($k = 1$) is normalized to 2π, then $\Delta x \Delta k = 2\pi/N$, and

$$\hat{F}_k = \sum_{j=0}^{N-1} f_j e^{-2\pi ikj/N}$$

Here the transform is written in the complex domain, but it could also be written in terms of trigonometric functions, as we will do below.

Fourier transforms can be calculated more quickly than one may have thought, with a method called the Fast Fourier Transform (FFT), a matter that will be revisited in chapter 10.

Polynomial interpolation was discussed in chapter 6. An alternative is to approximate a (real-valued) function $f(x)$ with a series of trigonometric functions,

$$\tilde{f}(x) = \frac{a_0}{2} + \sum_{k=1}^{\infty} a_k \cos(kx) + b_k \sin(kx)$$

We denote the Fourier representation of $f(x)$ with $\tilde{f}(x)$ in case the two cannot be made to agree exactly. (After all, countably many coefficients cannot be expected to provide a perfect match to a function everywhere, because a function defined on a continuous interval is made up of uncountably many points.)

Trigonometric functions obey an orthogonality relation:

$$\int_{-\pi}^{\pi} \cos(kx) \cos(mx)\, dx = \begin{cases} 2\pi & \text{for} \quad m = k = 0 \\ \pi & \text{for} \quad m = k \neq 0 \\ 0 & \text{if} \quad m \neq k \end{cases}$$

where k and m are integers. The equivalent orthogonality relation holds true for the sines, and the integrals of the cross-terms vanish, $\int_{-\pi}^{\pi} \cos(kx) \sin(mx)\, dx = 0$. When the above series representation for $\tilde{f}$ is multiplied with $\cos(mx)$ and integrated, we obtain, with the help of the orthogonality relations,

$$\int_{-\pi}^{\pi} f(x) \cos(kx)\, dx = \pi a_k$$

which provides an expression for the coefficients a_k. Likewise,

$$b_k = \frac{1}{\pi} \int_{-\pi}^{\pi} f(x) \sin(kx)\, dx$$

The coefficients of the discrete Fourier representation can be calculated with these integrals.

There are (at least) two perspectives on the Fourier coefficients. One is that the Fourier series exactly matches the function at equally spaced intervals. The other is that the Fourier coefficients minimize the square-deviation between the function and its approximation (the 2-norm). We will prove neither of these two properties (but see Exercise 7.3). Instead, we will analyze how Fourier approximations respond to discontinuities. Any nonperiodic function effectively represents a discontinuity, so a discontinuity is a rather common circumstance.

Consider a step-wise function $f(x) = 1$ for $|x| < \pi/2$ and $f(x) = 0$ for $|x| > \pi/2$ (Figure 7.1). According to the above formula, the Fourier coefficients are

$$a_k = \frac{1}{\pi} \int_{-\pi}^{\pi} f(x)\cos(kx)dx = \int_{-\pi/2}^{\pi/2} \cos(kx)dx = \frac{1}{\pi} \left. \frac{\sin(kx)}{k} \right|_{x=-\pi/2}^{\pi/2}$$

$$= \frac{2\sin\left(k\frac{\pi}{2}\right)}{k\pi}$$

and further

$$a_k = \begin{cases} \pi & \text{for } k = 0 \\ 0 & \text{for } k \text{ even} \\ 2\dfrac{(-1)^{(k+1)/2}}{k} & \text{for } k \text{ odd} \end{cases}$$

And $b_k = 0$, because $f(x)$ is even.

Figure 7.1 shows the function and the first 50 terms of its Fourier representation. It does not represent the function everywhere. By design, the Fourier approximation goes through every required point, uniformly spaced, but near the discontinuity it is a bad approximation in between grid points. This is known as the "Gibbs phenomenon", and occurs for approximations of discontinuous functions by Fourier series. This problem is similar to the "Runge phenomenon" that can occur for polynomials (Figure 6.2).

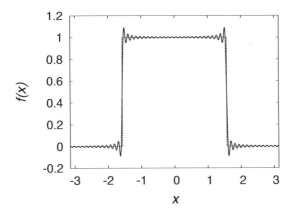

FIGURE 7.1 The Gibbs phenomenon. The Fourier series (solid line) of a discontinuous function (dot line) does not approximate the discontinuous function everywhere, no matter how many terms in the series are used.

In the aforementioned example of the step-function, the Fourier coefficients decay as $|a_k| \propto 1/k$. From the perspective of the Fourier series, a nonperiodic function has a step discontinuity at the boundary. Whenever that is the case, its coefficient will decay no slower than $1/k$. (Exercise 7.4 works this out in more detail.)

In chapter 6 the Weierstrass Approximation Theorem for polynomial approximations was stated. Here is one theorem for Fourier approximations (among the many known): When f is continuous, differentiable, and periodic, then the Fourier approximation converges uniformly. In other words, for such a function the maximum difference (the ∞-norm) between f and $\tilde{f}$ goes to zero as the number of Fourier terms increases. A continuous, differentiable, and periodic function is very well-behaved, and there are less-behaved functions where the Fourier approximation also works well, especially in the 2-norm, which it minimizes, but this theorem states that a function *that* nice can definitely be approximated uniformly. (By the way, a corollary of the theorem is that a continuous, differentiable, and periodic function *can* be described accurately with countably many coefficients.)

7.3 ORDINARY DIFFERENTIAL EQUATIONS

A common application is the numerical integration of ordinary differential equations (ODEs), $y'(t) = f(t, y)$, starting from an initial value $y(t = 0)$.

The simplest way to integrate an ODE is with a forward time difference, where the derivative is approximated by $dy/dt = (y^{n+1} - y^n)/\Delta t$; the superscripts number the time steps. The solver then proceeds as

$$y^{n+1} = y^n + f(t^n, y^n)\Delta t$$

Such a simple scheme for integrating differential equations is called "Euler method" or "Euler step". Although it has its own name, it is really the simplest conceivable method anyone would have come up with.

Needless to say, the Euler method is not the most accurate method, and we know from chapter 6 that the forward difference is only first-order accurate, so the error for y^{n+1} is $\Delta t \times O(\Delta t) = O(\Delta t^2)$ at each time step. Over a fixed interval of time, there are $O(1/\Delta t)$ steps, so the total error is $O(\Delta t)$.

Evaluating the derivative at half a time step leads to the following two-step scheme:

$$
\begin{aligned}
y^{n+\frac{1}{2}} &= y^n + \frac{1}{2}f(t^n, y^n)\Delta t \\
y^{n+1} &= y^n + f(t^{n+\frac{1}{2}}, y^{n+\frac{1}{2}})\Delta t
\end{aligned}
$$

This is called a "mid-point method". The second of these equations is no different from a center difference:

$$\frac{dy}{dt} = \frac{y^{n+\frac{1}{2}} - y^{n-\frac{1}{2}}}{\Delta t} = f(t^n, y^n)$$

shifted by half a time step. The mid-point method requires two function evaluations for the same time step, but the order of the error improves. The appropriate comparison is with two Euler steps with $\Delta t/2$ each. The error for the mid-point method is $O(\Delta t^2)$, a factor of Δt less than for one or two Euler steps. Vice versa, for the same accuracy a larger time step can be taken with the two-step method, so its computational efficiency is better than that of the Euler method. The mid-point method is also known as 2nd-order Runge-Kutta method.

Higher order methods can be constructed by using more function evaluations from the past. And lower order steps can be used to launch the higher order method, which would otherwise require values for the solution from before the start time.

Accuracy is not all we desire from an ODE integrator. An estimate of the discretization error is much desired, and a variable step size can greatly increase efficiency, plus it allows adjusting the step size to meet an accuracy goal. Error control is achieved by evaluating the outcome of the integration at two different resolutions, and that difference is used as proxy for the error. These error estimates can in turn be used to decide when it is time to increase or decrease the time step, something called "adaptive step size." And all this is ideally achieved at excellent computational efficiency, with a minimum number of function evaluations.

A most popular choice of a numerical integrator that efficiently combines all this desired functionality is the 4th order *Runge-Kutta method*. We will not study the details of these integrators, because there are plenty of excellent implementations around. Suffices to say that Runge-Kutta methods use finite difference approximations for the derivative and can adapt step size based on local error estimates. There are, of course, other types of integrators, and some ODEs require other methods, but 4th order Runge-Kutta is an excellent general purpose numerical integrator.

Nevertheless, the Euler method often becomes handy. For higher order methods, more arguments have to be passed to the integrator, along with any parameters f may depend on. Although straightforward to implement, this may require significant additional coding effort, plus additional validation effort. Overall, it may be less effort to use the simplest implementation and choose a small time step. Also, if f is not sufficiently smooth, then higher orders will not provide higher accuracy, because their error terms involve higher derivatives.

Now that basic thoughts about the numerical integration of ODEs have been expressed, we continue with a few comments that expand upon the numerics of ODEs in several directions.

A second order differential equation can be reduced to two first order differential equations by introducing a new variable $y_2 = y'$, where y can be thought of as y_1. The differential equation $y'' = f(t, y, y')$ then becomes $y_2' = f(t, y_1, y_2)$, and combined with $y_1' = y_2$, they make up two coupled first order ODEs. More generally, a high order differential equation can always be reduced to a system of coupled first order differential equations. Runge-Kutta methods and other common methods are readily generalized to systems of first-order equations by letting the derivative become multi-dimensional.

Numerical solution of some ODEs can be numerically unstable, in the same sense as the numerical instabilities of iterations described in chapter 2. An initial error, here caused by discretization error, is amplified and the numerical solution may rapidly grow away from the exact solution.

Consider $y'(t) = -ay(t)$ with $a > 0$ and $y(0) = 1$. The analytical solution is $y = \exp(-at)$. With a Euler step,

$$\frac{y^{n+1} - y^n}{h} = -ay^n$$

the time-stepping scheme is $y^{n+1} = y^n(1 - ah)$. Hence the time step better be $h < 1/a$, or the numerical solution will oscillate. For $h > 2/a$ it would even grow instead of decay. Hence, for this situation the Euler scheme is only "conditionally stable"; it is numerically stable for small time steps but numerically unstable for large time steps. With a backward difference

$$\frac{y^{n+1} - y^n}{h} = -ay^{n+1}$$

the scheme becomes $y^{n+1} = y^n/(1 + ah)$. This scheme works for any time step h, and is hence "unconditionally stable". For large time steps the solution will be inaccurate, but it does not oscillate or grow.

For $y' = -ay$, the solution changes on the time scale of $1/a$, so for accuracy reasons alone we would want to choose time steps smaller than that. *Stiff* differential equations involve two or more time-scales of very different scales, and integration based on the shortest time scale may no longer be practical. These stiff ODEs require special numerical methods that cure the instability.

Numerical integration always involves discretization errors. For differential equations that have conserved quantities we want to avoid that these errors accumulate steadily away from these conserved quantities. For example, for two-body gravitational interaction the energy is conserved and an orbit that would spiral in or out with time would betray the qualitative character of the solution, as seen in Figure 5.4(b). Special-purpose integrators can be designed with a discretization chosen to minimize the error of conserved quantities. In the case of gravitational interaction these are called "symplectic integrators" (where "symplectic" refers to the mathematical structure of the equations that describe Hamiltonian systems). Symplectic integrators are a special purpose method, but they illustrate a more general approach, namely that the freedom we have in choosing among various finite-difference approximations can be exploited to gain something more.

7.4   SYMBOLIC COMPUTATION

Computer Algebra Systems

Symbolic computation can reduce an expression such as ab/b^2 to a/b or evaluate the integral of x^2 as $x^3/3 + C$. This can be very time-saving. For example, integration of rational functions (ratios of two polynomials) would require one to sit down and work out a partial fraction decomposition—a tedious procedure. Instead, we can get answers promptly using symbolic computation software.

A sample session with the program Mathematica:

```
/* simple indefinite integral */

In[1]:= Integrate[x^2,x]

Out[1]= x^3/3 + constant
```

It even produced the integration constant.

```
/* integration of rational function */

In[2]:= Integrate[(4-x^2+x^5)/(1+x^2),x]

                2    4                              2
               x    x                          Log[1 + x ]
Out[2]= -x  -  -- + -- + 5 ArcTan[x]  +  -----------
               2    4                              2
```

```
/* roots of a 4th degree polynomial */

In[3]:= Solve[-48-80*x+20*x^3+3*x^4==0,x]

Out[3]= {{x -> -6}, {x -> -2}, {x -> -2/3}, {x -> 2}}
```

A 4th degree polynomial root and the integration of a rational function are in general very tedious to obtain by hand; both were mentioned in the Brainteaser of chapter 1.

The next example is the finite-difference approximation for the second derivative of a function:

```
/* Taylor expansion around 0 up to 5th order */

In[4]:= Series[f[h] - 2 f[0] + f[-h],{h,0,5}]

Out[4]= f''[0] h^2 + (1/12) f^(4)[0] h^4 + O[h]^6
```

```
/* infinite sum identified as Riemann zeta function */

In[5] = Sum[1/k^s,{k=1,Infinity}]

Out[5] = Zeta[s]
```

```
/* assumptions can be placed */
```

```
In[6]= Simplify[Sqrt[x^2], Assumptions -> x>0]
```

```
Out[6]= x
```

In Python, a module called SymPy offers symbolic algebra capabilities. For example,

```
>>> from sympy import *
>>> x = symbols('x')
>>> integrate(x**2, x)
```

```
x**3/3
```

```
>>> solve(-48-80*x+20*x**3+3*x**4, x)
```

```
[-6, -2, -2/3, 2]
```

```
# Taylor expansion around x=0 to 4th order
```

```
>>> series(sin(x)/x,x,0,4)
```

```
1 - x**2/6 + O(x**4)
```

This command not only produced a Taylor expansion to 4th order, it also realized that $\lim_{x \to 0} \sin(x)/x = 1$ for the first term.

```
# infinite sum
```

```
>>> k = symbols('k')
>>> summation(k**(-4), (k,1,oo))
```

```
pi**4/90
```

That double-oh symbol indeeds stands for ∞, and the command symbolically obtained the sum of the infinite series.

Symbolic computation software is efficient for certain tasks, but cannot compete with the human mind for others. For example, simplifying expressions is something humans are still often better in than computers. The human brain recognizes simplifications that elude symbolic algebra packages, especially when the same calculation is carried out by hand over and over again. And if a symbolic computation program does not symbolically integrate an expression, it does not entirely exclude that an explicit formula does not exists.

Software packages can have bugs and mathematical handbooks can have typos. Fortunately, bugs tend to get corrected over time, in books as well as in software. By now symbolic computation software has matured in many ways and is rather reliable.

Currently available comprehensive packages with symbolic computation abilities include *Axiom, Maple, Mathematica, Maxima, Reduce*, and (Python-based) *SymPy*. A free interface to one-line Mathematica commands is provided at `www.wolframalpha.com`.

Diagrammatic perturbation expansion*

One kind of symbolic computation, used by humans and for computers, are diagrammatic perturbation expansions that help us keep track of the many terms and multipliers a power series expansion can have.

Consider a classical gas at temperature T. The partition function, which is key to calculating thermal properties of a statistical system, is the sum of exponential factors $\exp(-E/kT)$, where E is kinetic plus potential energy and k the Boltzmann constant. Denote $\beta = 1/kT$ for brevity. The kinetic energy of a particle is its momentum squared divided by two times its mass, $p^2/2m$. The potential energy is given by the distance r between particles via the pairwise potential $u(|r|)$, whose form we keep general.

For one particle there is no potential energy and the partition function is

$$Z(\beta, V, m) = C \int dp \int dr e^{-\frac{\beta p^2}{2m}}$$

where V is the volume and C a physical constant. (That constant contains information about how many elements the sum has within a continuous interval, something that comes from deeper physics.) For two particles the partition function is

$$Z(\beta, V, m) = C \int dp_1 dp_2 \int dr_1 dr_2 e^{-\beta\left[\frac{p_1^2}{2m} + \frac{p_2^2}{2m} + u(r_1 - r_2)\right]}.$$

The integrals over momentum could be carried out since they are Gaussian integrals. However, a gas without interactions is an ideal gas and hence we simply write

$$Z = Z_{\text{ideal}} \int dr_1 dr_2 e^{-\beta u(r_1 - r_2)}$$

which also absorbs the constant C. As we all know, the series expansion of an exponential is $e^x = 1 + x + x^2/2! + x^3/3! + ...$. An expansion for small β, that is, high temperature, yields

$$Z = Z_{\text{ideal}} \int dr_1 dr_2 [1 - \beta u_{12} + \frac{\beta^2}{2} u_{12}^2 + ...]$$

Here, we abbreviated $u_{12} = u(|r_1 - r_2|)$. For three particles,

$$\frac{Z}{Z_{\text{ideal}}} = \int dr_1 dr_2 dr_3 e^{-\beta(u_{12}+u_{13}+u_{23})}$$

$$= \int dr_1 dr_2 dr_3 [1 - \beta(u_{12} + u_{13} + u_{23}) +$$

$$+ \frac{\beta^2}{2!}(u_{12}^2 + u_{13}^2 + u_{23}^2 + 2u_{12}u_{13} + 2u_{12}u_{23} + 2u_{13}u_{23}) + ...]$$

Many of these integrals yield identical results (for $V \to \infty$), so we arrive at

$$\frac{Z}{Z_{\text{ideal}}} = \int dr_1 dr_2 dr_3 \left[1 - 3\beta u_{12} + 3\frac{\beta^2}{2!}(u_{12}^2 + 2u_{12}u_{13}) + ...\right]$$

To keep track of the terms, they can be represented by diagrams. For example,

$$\int dr_1 dr_2 dr_3 u_{12} =$$

Since u_{12}, u_{13}, and u_{23} yield the same contribution, one diagram suffices to represent all three terms. The full perturbation expansion for Z/Z_{ideal} in diagrammatic form is

The number of dots is the number of particles. The number of lines corresponds to the power of β, that is, the order of the expansion. Every diagram has a multiplication factor corresponding to the number of distinct possibilities it can be drawn. In addition, it has a factorial in the denominator (from the coefficients in the series expansion of the exponential function) and a binomial prefactor (from the power).

Finally, unconnected dots can be omitted, as all required information is already contained otherwise. The diagrammatic representation of the expansion simplifies to

Using the diagrams, it is straightforward to write down the perturbation expansion for more particles or to higher order (Exercise 7.8).

Not every diagram requires to calculate a new integral. For example, the four-particle term $\int dr_1 dr_2 dr_3 dr_4 u_{12} u_{34} = \left(\int dr_1 dr_2 u_{12}\right)\left(\int dr_3 dr_4 u_{34}\right)$. Hence, this diagram can be reduced to the product of simpler diagrams:

Disconnected parts of diagrams always multiply each other.

The high-temperature expansion of the partition function in statistical mechanics is known as "(Mayer) cluster expansion." In quantum field theory, perturbation expansions are extensively used in the form of "Feynman diagrams." Elaborate symbolic computation programs have been written to obtain such perturbation expansions to high order.

EXERCISES

7.1 Derive the equations for linear regression without intercept, $y = kx$, i.e., given data pairs (x_i, y_i), $i = 1, ..., N$, find the equation for k that minimizes the sum of quadratic deviations, $\sum_{i=1}^{N} (y - y_i)^2$.

7.2 Weighting proportional with distance improves the robustness of a fit. This method minimizes $\sum_i |y_i - a - bx_i|$, where x_i and y_i are the data and a and b are, respectively, the intercept and slope of a straight line. Show that finding the parameters a and b requires nonlinear root-finding in only one variable.

7.3 Show that the Fourier coefficients minimize the square-deviation between the function and its approximation. Use the trigonometric representation.

7.4 a. Calculate the Fourier coefficients of the continuous but nonperiodic function $f(x) = x/\pi$, on the interval $[-\pi, \pi]$. Show that their absolute values are proportional to $1/k$.

   b. Any nonperiodic continuous function can be written as the sum of such a ramp and a periodic function. From this, argue that continuous nonperiodic functions have Fourier coefficients that decay as $1/k$.

   c. Calculate the Fourier coefficients of a tent-shaped function, $f(x) = 1 - |x|/\pi$ on $[-\pi, \pi]$, which is continuous and periodic, and show that their absolute values are proportional to $1/k^2$.

7.5 Consider a numerical solver for the differential equation $y' = -ay$ that uses the average of a forward time step and a backward time step:

$$\frac{y^{n+1} - y^n}{h} = -a\frac{y^n + y^{n+1}}{2}$$

   a. Derive the criterion for which the numerical solution will not oscillate.

   b. Show that this method is one order more accurate than either the forward- or the backward-step method.

7.6 Using a symbolic computation software of your choice, calculate the 10th derivative of $1/(1 + 25x^2)$ at $x = 0$. (This function was used in Figure 6.2.) The main purpose of this exercise is to learn how to use a computer algebra system and some of its syntax.

7.7 a. Using a symbolic computation software of your choice, find the roots of $x^5 - 7x^4 + 2x^3 + 2x^2 - 7x + 1 = 0$. (This equation was used in the Brainteaser of chapter 1.)

   b. With or without the assistance of software, verify that the answers are correct.

7.8 For the diagrammatic expansion method for $e^{-\beta u}$, described in section 7.4:

   a. Write down all distinct diagrams up to second order in β with N particles.

   b. Determine the prefactor for each of the terms.

## CHAPTER 8

Performance Basics & Computer Architectures

This chapter marks the beginning of the portion of the book that is primarily relevant to "large" computations, for which computational performance matters, because of the sheer number of arithmetic operations, the rate at which data have to be transferred, or anything else that is limited by the physical capacity of the hardware.

8.1 EXECUTION SPEED AND LIMITING FACTORS OF COMPUTATIONS

Basic floating-point operations, such as addition and multiplication, are carried out directly on the central processing unit (CPU). Elementary mathematical functions, such as exponential and trigonometric functions, on the other hand, are emulated on a higher level. Table 8.1 provides an overview of the typical relative execution times for common mathematical operations.

The relative speeds in Table 8.1 are much unlike doing such calculations with paper-and-pencil. On a computer, additions are no faster than multiplications, unlike when done by hand. On modern processors multiplication takes hardly longer than addition and subtraction. Divisions take longer than additions or multiplications, both for integers and floating-point numbers. Replacing a division by a multiplication saves time.

As a historical note, there was a time when multiplications took considerably longer than additions on a CPU. At that time, it was appropriate to worry more about the number of multiplications than about the number of additions, and if a reformulation of an equation would replace a multiplication by several additions, it was favorable for speed, but this is no longer the case.

A unit of 1 in Table 8.1 corresponded to about 1 nanosecond on a personal computer or workstation (with a clock frequency of about 3 GHz). Contemplate for a moment how many multiplications can be done in one second:

TABLE 8.1 The approximate relative speed of integer operations, floating-point operations, and several mathematical functions. (The results in this table are based on C and Fortran programs compiled with various compilers and executed on various platforms. "Double" refers to a double-precision floating-point number.)

arithmetic operation	relative time	clock cycles
integer addition or subtraction	≤1	1
integer multiplication	≤1	1
integer division	2–5	7–9
double addition or subtraction	1	3
double multiplication	1	3–5
double division	4–10	10–20
sqrt	7–20	20–23
sin, cos, tan, log, exp	25–100	60–250

10^9—a *billion* floating-point operations per second (FLOPS). (The "S" stands for "per second" and is capitalized, whereas FLOPs is the plural of FLOP.)

We can verify that these execution speeds are indeed reality. A simple interactive exercise with a program that adds a number a billion times to a variable can be used to approximately measure absolute execution times. Such an exercise also reveals the ratios listed in the table above. We simply write a big loop (according to chapter 3, a 4-byte integer goes up to about 2×10^9, so a counter up to 10^9 does not overflow yet), compile the program, if it needs to be compiled, and measure the runtime, here with the Unix command `time`, which outputs the seconds consumed.

```
> gfortran speed.f90       % 10^9 additions
> time a.out
2.468u 0.000s 0:02.46
```

One billion additions in 2.46 seconds demonstrates that our program indeed processes at a speed on the order of one GigaFLOPS. If optimization is turned on during compilation, execution is about three times faster in this example, achieving a billion additions in less than a second:

```
> gfortran -O speed.f90    % 10^9 additions with optimization
> time a.out
0.936u 0.000s 0:00.93
```

While we are at it, we can also check how fast Python is at this. Here we only loop up to 10^8 instead of 10^9; otherwise, we would wait for a while.

```
> time python speed.py     % 10^8 additions
6.638u 0.700s 0:07.07
```

The result reveals that Python is far slower than Fortran, as already stated in chapter 4.

In terms of standard numerical tasks, one second is enough time to solve a linear system with thousands of variables or to take the Fourier Transform of a million points. In conclusion:

Arithmetic is fast!

In fact, arithmetic is so fast that the bottleneck for numerical computations often lies elsewhere. In particular, for a processor to compute that fast, it also needs to be fed numbers fast enough.

There are basically four limiting physical factors to computations:

- Processor speed
- Memory size
- Data transfer between memory and processor
- Input and output

We proceed by visiting each of the remaining factors.

8.2   MEMORY AND DATA TRANSFER

Memory size

The memory occupied by a number can be language implementation and machine dependent, but standardization has led to significant uniformity. In most programming languages, each data type takes up a fixed number of bytes, and these allocations most commonly are

- 4 bytes for an integer,
- 4 bytes for a single-precision number, and
- 8 bytes for a double-precision number.

Hence the required memory can be precisely calculated. For example, an array of 1024×1024 double-precision numbers takes up *exactly* eight Mebibytes. A Mebibyte is 2^{20} bytes which is only approximately 10^6 bytes (1 Megabyte). Similarly, a kibibyte (abbreviated as KiB or KB) is 1024 bytes, not 1000 bytes (1 kilobyte, kB). The same goes for Gibi and Giga.

In some languages all variables need to be declared. For example, in C, int defines an integer, but it does not guarantee a specific byte length for it. That said, it will most likely be a 32-bit (4-byte) integer. To prescribe a specific byte-length, special declarations need to be used. In C, int32_t is guaranteed to be 32 bits, whereas the byte-length of int is dependent on the machine.

In other programming languages variable types are automatic. For example in Matlab/Octave every variable is by default a double-precision floating-point number, even integers are. There is, however, the option to declare a variable type differently, for example, shorter to save on the memory size of an array. Some languages have an intrinsic 1-byte data type, called "boolean" or "logical", that stores true/false, which saves memory compared to an integer.

A true-false flag only requires one bit, not one byte, so it still consumes more memory than necessary.

According to the above rule, the total memory consumed is simply the memory occupied by the variable type multiplied by the number of elements. There is an exception to this rule for calculating required memory. For a compound data type, and many languages offer compound data types, the alignment of data can matter. If, in C notation, a data type is defined as struct {int a; double b}, it might result in unused memory. Instead of 4+8 bytes, it may consume 8+8 bytes. Compilers choose such alignments in compound data types to make memory accesses faster, even though it can result in unused memory.

Programs can access more (virtual) memory than physically exists on the computer. If the data exceed the available memory, the hard drive is used for temporary storage (swap space). Reading and writing from and to a hard drive is orders of magnitude slower than reading from memory, so this slows down the calculation dramatically. If the computer runs out of physical memory, it will try to place those parts in swap space that are not actively used, but if the CPU needs data that are not in physical memory, so it has to fetch them from disk, it will sit idle for many, many clock cycles until those data arrive. Memory is organized in blocks of fixed length, called "pages." If the data do not exist in memory, then the page needs to be copied into memory, which delays execution in the CPU.

Data transfer

Data transfer between memory and processor. When a processor carries out instructions it first needs to fetch necessary data from the memory. This is a slow process, compared to the speed with which the processor is able to compute. This situation is known as the "processor-memory performance gap".

A *register* is a memory unit located on the central processing unit that can be accessed promptly, within one clock cycle. Registers are a very small and very fast storage unit on which the processor operates.

To speed up the transport of data a "cache" (pronounced "cash") is used, which is a small unit of fast memory. Frequently used data are stored in the cache to be quickly accessible for the processor. Data are moved from main memory to the cache not byte by byte but in larger units of "cache lines," assuming that nearby memory entries are likely to be needed by the processor soon (assumption of "spatial locality"). Similarly, "temporal locality" assumes that if a data location is referenced then it will tend to be referenced again soon. If the processor requires data not yet in the cache, a "cache miss" occurs, which leads to a time delay.

A hierarchy of several levels of caches is customary. Table 8.2 provides an overview of the memory hierarchy and the relative speed of its components. The large storage media are slow to access. The small memory units are fast. (Only one level of cache is included in this table. Higher levels of cache have

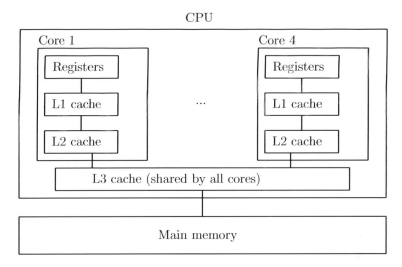

FIGURE 8.1 Schematic of memory hierarchy in Intel Core i7.

access times in between that of the level-1 cache and main memory.) Figure 8.1
illustrates the physical location of memory units within a multi-core CPU.

TABLE 8.2 Memory hierarchy and typical relative access times (in units
of clock cycles).

Registers	1
Level 1 Cache	4
Main memory	120
Magnetic disk	10^7

Data transfer to and from hard drive. Currently there are two common
types of hard drives: solid state drives (SSD) and magnetic drives, also known
as hard disk drives (HDD). Magnetic drives involve spinning disks and mechanical arms. Non-volatile flash (solid-state) memory is faster (and more
expensive per byte) than magnetic drives, and slowly wears out over time.

The time to access the data on a hard drive consists of seek time (latency)
and data transfer time (bandwidth). Reading or writing to or from a magnetic
disk takes as long as millions of floating-point operations. The majority of this
time is for the head, which reads and writes the data, to find and move to
the location on the disk where the data are stored. Consequently data should
be read and written in big chunks rather than in small pieces. In fact the
computer will try to do so automatically. While a program is executing, data
written to a file may not appear immediately. The data are not flushed to the
disk until they exceed a certain size or until the file is closed.

Table 8.3 shows the access times (latency) for main memory and both types of hard drives. These times are best understood in comparison to CPU speed, already discussed above.

TABLE 8.3 Execution and access times for basic computer components. Most of these numbers are cited from Patterson & Hennessy (2013).

CPU		1 ns
Main Memory		50–70 ns
Hard drive	Solid-state	0.1 ms $= 10^5$ns
	Magnetic disk	5–10 ms $= 10^7$ns

Input/output to other media. Table 8.3 reveals how critical the issue of data transfer is, both between processor and memory and between memory and hard disk. Input and output are relatively slow on any medium (magnetic harddisk, display, network, etc.). Writing on the display is a particularly slow process; excesses thereof can easily delay the program. A common beginner's mistake is to display vast amounts of output on the display, so that data scroll down the screen at high speed; this slows down the calculation (Exercise 8.3).

Input/output can be a limiting factor due to the data transfer rate, but also due to sheer size. Such problems are data-intensive instead of compute-intensive.

Table 8.4 shows an overview of potential resource constraints. The current chapter deals with compute-intensive problems. Data-intensive problems will be dealt with in chapters 13 and 14. When neither is a significant constraint, then the task of solving a computational problem is limited by the time it takes a human to write a reliable program. Most everyday scientific computing tasks are of this kind.

TABLE 8.4 Overview of physical resource constraints

Compute intensive	arithmetic operations memory-CPU data transfer bandwidth memory size
Data intensive	storage size hard drive-memory data transfer bandwidth network bandwidth

8.3 A PROGRAMMER'S VIEW OF COMPUTER HARDWARE

In this section we look at how a program is processed by the computer and follow it from the source code down to the level of individual bits executed on the hardware.

The lines of written program code are ultimately translated into hardware-dependent "machine code". For instance, the following simple line of code adds two variables: a=i+j. Suppose i and j have earlier been assigned values and are stored in memory. At a lower level we can look at the program in terms of its "assembly language," which is a symbolic representation of the binary sequences the program is translated into:

```
lw $8, i
lw $9, j
add $10, $8, $9
sw $10, a
```

The values are pulled from main memory to a small memory unit on the processor, called "register," and then the addition takes place. In this example, the first line loads variable i into register 8. The second line loads variable j into register 9. The next line adds the contents of registers 8 and 9 and stores the result in register 10. The last line copies the content of register 10 to variable a, that is, its memory address. There are typically about 32 registers; they store only a few hundred bytes. Arithmetic operations, in fact most instructions, only operate on entries in the registers. Data are transferred from main memory to the registers, and the results of operations written back out to memory. A hardware-imposed universal of numerical computations is that operations have no more than two arguments, each of fixed bit length.

At the level of the assembly language or machine code there is no distinction between data of different types. Floating-point numbers, integers, characters, and so on are all represented as binary sequences. What number actually corresponds to the sequence is a matter of how it is interpreted by the instruction. There is a different addition operation for integers and floats, for example. (This sheds light on what happens if a variable of the wrong type is passed to a function. Its bit pattern ends up not only slightly misinterpreted, but completely misinterpreted.)

Instructions themselves, like lw and add, are also encoded as binary sequences. The meaning of these sequences is hardware-encoded on the processor. These constitute the "instruction set" of the processor. (An example is the "x86" instruction set, which is commonly found on computers today.) When a program is started, it is first loaded into memory. Then the instructions are executed.

During consecutive clock cycles the processor needs to fetch the instruction, read the registers, perform the operation, and write to the register. Depending on the actual hardware these steps may be split up into even more substeps. The idea of "pipelining" is to execute every step on a different, dedicated element of the hardware. The next instruction is already fetched, while the previous instruction is at the stage of reading registers, and so on. Effectively, several instructions are processed simultaneously. Hence, even a single processor core tries to execute tasks in parallel, an example of instruction-level parallelism.

The program stalls when the next instruction depends on the outcome of the previous one, as for a conditional statement. Although an `if` instruction itself is no slower than other elementary operations, it can stall the program in this way. In addition, an unexpected jump in the program can lead to cache misses. For the sake of computational speed, programs should have a predictable data flow.

The program in Table 8.5 illustrates one of the practical consequences of pipelines. The dependence of a variable on the previous step leads to stalling. Hence, additions can be accomplished at even higher speed than shown in Table 8.1, if subsequent operations do not immediately require the result of the addition. (The execution speed of operations that require only one or a small number of clock cycles effectively depends on the sequence of commands they are embedded in.)

TABLE 8.5 Although both of these Fortran loops involve two billion additions, the version to the right is twice as fast.

```
do i=1,1000000000          do i=1,1000000000
   a=a+12.1                    a=a+12.1
   a=a+7.8                     b=b+7.8
end do                     end do
```

A computer's CPU is extremely complex. For instance, constants defined in a program may be stored in the register. This decision is automatically made by the compiler. The processor uses its own intelligence to decide what data to store in a register. And for conditional operations, it may speculate which of the possible branches is most likely to occur next. Executing the correctly predicted branch of an if statement is faster than executing the "wrong" (mispredicted) branch of an if statement. Much of this complexity is hidden from the user, even at the assembly language level, and taken care of automatically.

To revisit parts of what we learned so far, Table 8.6 shows the results of a small case study for the execution time of variants of floating-point addition and the square root. Simple addition of double-precision numbers took less than one nanosecond on this particular CPU, or about 3 clock cycles. The absolute execution time is subject to change with technological advances, but the number of cycles per operation has varied far less over time. "Concurrent" refers to operations that can be carried out independently from one another, as the right-hand side example in Table 8.5. In this case, up to three additions consumed no more time than a single addition would; only a fourth concurrent addition increased the total time. For many concurrent additions, the throughput is one addition per cycle, thanks to pipelining, so these simple operations are effectively processed at the speed of the clock cycle itself.

According to the third row in the table, a single-precision float addition is slower than a double-precision addition (as mentioned in chapter 3), because of conversions to and from the processor's intrinsic representation. For mathematical operations that require many clock cycles, such as the square root, float is faster than double, because the accuracy of the result can be lower.

TABLE 8.6 Case study for execution times on a CPU with 3.6 GHz clock frequency and the gcc compiler.

operation	time per operation (ns)	clock cycles
double addition	0.91	3
concurrent double addition	0.26	1
float addition	2.5	9
double sqrt	6.5	23
float sqrt	4.5	16

8.4 COMPUTER ARCHITECTURES AND TECHNOLOGICAL CHANGE

All modern CPUs have multiple cores. These cores are independent processors on a single chip. A multi-core CPU is essentially a multi-CPU system, only faster. Currently, four or six-core CPUs are the most common, but even 18-core CPUs are readily available, although with a lower clock rate. This represents a significant parallelism that every user can take advantage of.

At the coarse level, the hardware architecture of a computer is the following: A CPU core contains control and arithmetic units. Instructions operate on binary numbers or binary memory addresses stored in registers. The main memory is physically separate from the CPU. The two are connected via several levels of caches, with the lowest one or two levels physically inside the CPU. All modern CPUs contain several cores. Displays, hard drives, and network are connected with the CPU and its memory at much slower bandwidth than exists between the CPU and main memory.

Pipelines operate within each CPU core. Another form of parallelism already present on a single core is multithreading. Vaguely speaking, a "thread" is the smallest sequence of instruction that can be programmed. Typically a CPU core has two threads, which improves the throughput. The creation and scheduling of these threads is automatically handled at a low level, but in some languages threads can be dealt with individually, if desired.

The mind-boggling speed of modern CPUs is related to the fabled "Moore's law". Moore's law is a statement about how the number of transistors per area increased with time. Combined with other improvements, it has translated into a dramatic increase in floating-point operations per second (FLOPS) over many decades. Over about half a century, until about the mid-2000's, processor performance doubled every 18–24 months. That amounts to one and a half to two orders of magnitude per decade. Since then, the rate of increase for

FLOPS per processor core has slowed, causing a sea change toward multi-core processors. This ushered in a new era for parallel computing. As someone put it: "Computers no longer get faster, just wider."

The ridiculously fast multi-core CPU is where essentially all scientific computations take place. Hence, for tasks where performance matters, the extremely fast multi-core CPU should be what we design our programs and algorithms for.

Computational performance is rooted in physical hardware, and the optimal numerical method therefore may depend on the technology of the time. Over the decades the bottlenecks for programs have changed with technology. Long ago it was memory size. Memory was expensive compared to floating-point operations, and algorithms tried to save every byte of memory possible. Later, memory became comparatively cheap, and the bottleneck moved to floating-point operations. For example, storing repeatedly used results in a temporary variable rather than recalculating them each time increased speed. This paradigm of programming is the basis of classical numerical analysis, with algorithms designed to minimize the number of floating-point operations. Today the most severe bottleneck often is moving data from memory to the processor. One can also argue that the bottleneck is the parallelizability of an algorithm.

Recommended Reading: Patterson & Hennessy's *Computer Organization and Design: The Hardware/Software Interface* is a very accessible textbook for this rapidly changing field by two eminent scientists. Updated editions are being published regularly. (Incidentally John Hennessy became the president of Stanford University, which did not stop him from working on updated editions of this book.) The book is for an undergraduate computer science level, and not to be confused with a graduate level textbook by the same two authors.

EXERCISES

8.1 Consider the dot product of two vectors

$$c = \sum_{i=1}^{N} x_i y_i$$

    a. If x and y are 64-bit variables and $N = 1000$ how many bytes of memory are consumed by x and y combined?

    b. How many floating-point operations are necessary to form the dot product?

    c. What is the ratio of data transferred from memory to processor to the number of floating-point operations, in units of byte per FLOP?

8.2 To find the minmum pairwise distance between a large number of points with coordinates (x, y), we can calculate the distances $\sqrt{(x_i - x_j)^2 + (y_i - y_j)^2}$ and then take the minimum. But because the square root is an expensive function, it is faster to only calculate the square of the distances and take the square root only once the minimum is found. In a programming language of your choice:

a. implement both versions,
b. find a way to measure the execution time, and
c. determine by what factor the computation time is reduced.

8.3 Slowdown from excessive output to display: Take a simple computation, such as $x_{n+1} = 4x_n(1 - x_n)$, and compute the first ten million iterations with $x_0 = 0.8$.

a. Output the result of each iteration to the display and measure the execution time. Then output only the last iteration and measure the execution time. Since both runs executed the same number of FLOPs, the difference in runtime is due to output to the display.
b. Instead of displaying the output, write it to a file. How does the execution time compare to the other two cases?

8.4 a. One way or another, find out the following hardware specifications of your computer: the number of CPU cores, the total amount of memory, and the memory of each level of cache.
b. Using a programming language of your choice, determine the number of bytes consumed by a typical floating-point number and a default integer. In each language, a dedicated command is available that returns this value.

## CHAPTER 9

High-Performance & Parallel Computing

9.1 CODE OPTIMIZATION

This section is about coding (code writing) with computational performance in mind. To use resources efficiently, the time saved through optimizing code has to be weighed against the human resources required to implement these optimizations. Depending on the application, these optimizations are worthwhile or not, and there certainly are applications where optimization truly matters. And sometimes writing well-performing code is as simple, or even simpler, than writing badly-performing code.

Optimization for memory size, data transfer, and arithmetic operations

In chapter 8 it has already been explained that a dramatic slowdown occurs when physical memory is exceeded and that data transfer is often a bottleneck in numerical calculations.

The basic storage unit for memory is 1 byte, and modern systems are "byte-accessible", meaning there is one memory address for each byte. Memory addresses are numbered in a linear manner. Even when an array has two or more indices, its elements are ultimately stored in physical memory in a one-dimensional fashion. Languages store array data with two indices either row-wise or column-wise. The fastest index of an array for Fortran, NumPy, and IDL is the first (leftmost) index and for C and C++ the last (rightmost) index. Although this is not part of the language standard, it is the universally accepted practice of compiler writers. Reading data along any other index requires jumping between distant memory locations, leading to cache misses. This is a perfect example of a performance-minded coding practice that requires no extra work; one just has to be aware of the fact (Figure 9.1).

As eluded to in chapter 8, data flow should be predictable to avoid costly cache misses. Loops should have a small stride to avoid cache misses. Condi-

low performance	high performance
| ```
do i=1,N
  do j=1,N
    freq2(i,j)=sqrt(1.*j**2+i**2)
  enddo
enddo
a(:,:) = freq2(:,:)*z(:,:)
``` | ```
do j=1,N
  do i=1,N
    freq2=sqrt(1.*j**2+i**2)
    a(i,j)=freq2*z(i,j)
  enddo
enddo
``` |

FIGURE 9.1 The Fortran example to the left accesses the first index with a stride. It also wastes memory by creating an unnecessary array for `freq2`, and, worse, this array first needs to be written to memory and then read in. All these issues are corrected in the version on the right side, where the inner loop is over the first index and data locality is observed.

tional statements, like `if`, should be avoided in the loop which does most of the work. The same holds true for statements that cause a function or program to intentionally terminate prematurely. Error, stop, or exit statements often—not always, but often—, demand that the program exits at *exactly* this point, which prevents the CPU from executing lines further down ahead of time that are otherwise independent. Such conditional stops can cause a performance penalty, even when they are never executed.

Obviously, large data should not be unnecessarily duplicated, and with a bit of vigilance this can be avoided. When an array is passed to a function or subroutine, what is passed is the address of the first element, so this does not create a duplicate array.

"Memoization" refers to the storing of intermediate results of expensive calculations, so they can be reused when needed. There can be a conflict between minimizing the number of arithmetic operations, total memory use, and data motion. For example, instead of storing intermediate variables for reuse, it may be better to re-compute them whenever needed, because arithmetic is fast compared to unanticipated memory accesses. It is often faster to re-compute than to re-use.

Program iteratively, not recursively. Rather than write functions or subroutines that call themselves recursively, it is more efficient to use iteration loops. Every function call has an overhead, and every time the function is called the same local variables need to be allocated.

Intrinsic functions can be expected to be highly efficient. Predefined vector reduction operations such as sum and dot product intrinsically allow for a high degree of parallelization. For example, if a language has a dedicated command for matrix multiplication, it may be highly optimized for this task. For the sake of performance, intrinsics should be used when available.

Loops can be treacherously slow in very high-level languages such as Python and Matlab. The following loop with a NumPy array

```
for i in range(32000):
    if a[i] == 0:
        a[i] = -9999
```

is faster and more conveniently implemented as

```
a[a == 0] = -9999
```

This is also known as "loop vectorization".

As mentioned previously, CPUs are extremely complex and often a naive guess about what improves or reduces computational efficiency is wrong. Only measuring the time of actual program execution can verify a performance gain.

Optimization by the compiler

Almost any compiler provides options for automatic speed optimization. A speedup by a factor of five is not unusual, and is very nice since it requires no work on our part. (In the example that was shown in section 8.1, a speedup of three resulted from turning on compiler optimization). In the optimization process the compiler might decide to get rid of unused variables, pull out common subexpressions from loops, rearrange formulae to reduce the number of required arithmetic operations, inline short functions, rearrange the order of a few lines of code, insert instructions within loops that prefetch data from main memory to avoid cache misses, and more. The compiler optimization cannot be expected to have too large of a scope; it mainly does local optimizations.

Rearrangements in equations can spoil roundoff and overflow behavior. If an expression is potentially sensitive to roundoff or prone to overflow, setting otherwise redundant parentheses might help. Some compilers honor them.

Compilers may overwrite some language constructs. For example, in C there is a keyword `register` to indicate that a variable should be placed in the register. The compiler may overwrite this, as it does its own (most likely better) optimization.

In a conditional statement with two arguments, say if (a>0 && b>0), we have no control over whether the first or second condition will be evaluated first. It could be either. This is not even due to compiler optimization, but the nature of program processing in general. For example, suppose we have an array with five elements p[0],...,p[4] and an integer i that takes values from 0 to 9. Then if (i<5 && p[i]>0) is a programming error, because the condition to the right might be evaluated a nanosecond before the condition to the left, which means an element of the array p is sought that does not exist.

At the time of compilation it is not clear (to the computer) which parts of the program are most heavily used or what kind of situations will arise. The optimizer may try to accelerate execution of the code for all possible inputs and program branches, which may not be the best possible speedup for the actual input. This is the one reason why Just-in-time compilers, mentioned in chapter 4, can in principle provide better optimization results than static compilers.

A compiler may offer fine-grained and detailed options for the functions it does or does not perform. For example, the gcc compiler has hundreds of options, some dedicated to specific CPU architectures. These are many more options than we would ever want to acquaint ourselves with, but some are worth pointing out. There are levels of aggressiveness for speed optimization, for example, -O0 to -O3. The zeroth level does no speed optimization at all and therefore compiles the fastest, while the highest level performs aggressive speed optimization. There can also be a conflict between aggressive speed optimization and the IEEE roundoff standard (Chapter 3). Additional options, that may be called "fast-math" or "unsafe-math-optimizations", provide control over whether or not exact roundoff requirements should be sacrificed for speed.

Another trick for speedup is to try different compilers and compare the speed of the executable they produce. Different compilers sometimes see a program in different ways. Compilers have been written by the open-source community, and most notable among these creations is the Gnu Compiler Collection; compilers have also been developed by several companies who sell them or make them available for free. When new CPU architectures or program standards appear, compilers have to be updated to exploit the new features.

Over time, compilers have become more intelligent and mature, taking away some of the work programmers needed to do to improve performance of the code. The compiler most likely understands more about the details of the computer's central processor than we do, but the programmer understands her code better overall and can assist the compiler in the optimization.

Performance analysis tools

Hidden or unexpected bottlenecks can be identified and improvements in code performance can be verified by measuring the execution time for the entire run or the fraction of time spent in each subroutine. The simplest measure is the execution time of the entire run, for example with the Unix command time. Even for a single user the execution time of a program varies due to other ongoing processes, say by 5%. Thus, a measured improvement by a few percent might merely be a statistical fluctuation.

Every common programming language has commands that can time events within the program. (Among the more entertaining are the Matlab commands tic, which starts a stopwatch, and toc, which determines the elapsed time.) Temporarily placing such timing commands into the code makes it possible to measure the execution time of individual components.

A variety of standard tools are available for fine-grained analysis. *Profilers* collect information while the program is executing. Profiling analysis reveals how much time a program spent in each function, and how many times that function was called. It can even count how many times each line in the source code was executed. If we want to know which functions consume most of the cycles or identify the performance bottleneck, this is the way to find out. The program may need to be compiled and linked with special options to enable detailed profiling during program execution. And because the program is all mangled up by the time it is machine code, those tracebacks can alter its performance, which is why these options are off by default. Some profilers start with the program; others can be used to inspect and sample any running process even when that process is already running. Table 9.1 shows a sample of the information a standard profiling tool provides.

TABLE 9.1   A fraction of the output of the profiling tool gprof. According to these results, most of the time is spent in the subroutine conductionq and the function that is called most often is flux_noatm.

% time	cumulative seconds	self seconds	calls	name
51.20	7.59	7.59	8892000	conductionq_
44.66	14.22	6.62	8892000	tridag_
2.02	14.52	0.30	9781202	flux_noatm_
1.82	14.79	0.27	9336601	generalorbit_
...				

9.2   PARALLEL COMPUTING

Modern CPUs are intrinsically parallel. In fact, purely serial CPUs are no longer manufactured. Hardware parallelism comes in several forms. In chapter 8 we have already encountered single-core parallelism in the form of pipelining. Here we focus primarily on the form of parallelism that is most scalable, namely multi-core and multi-CPU parallelism. While parallel hardware is readily available, writing programs that take proper advantage of that parallelism is often where the bottleneck lies. Fully exploiting parallel hardware can be a formidable programming challenge.

We begin with the concept of "concurrency." Concurrency refers to processes or tasks that can be executed in any order or, optionally, simultaneously. Concurrent tasks may be executed in parallel or serially, and no interdependence is allowed. Concurrency allows processors to freely schedule tasks to optimize throughput. The proper nomenclature is "serial" versus "parallel" for hardware, and "sequential" versus "concurrent" for programs, but the term "parallel" is also used in a more general sense.

A CPU with several processor cores and, even more so, a computer cluster with many CPUs can execute instructions in parallel. The memory can either still be shared among processors or also be split among processors or small groups of processors ("shared memory" versus "distributed memory" systems). An example of the former is a multi-core processor; the cores share the same main memory. For large clusters only distributed memory is possible. With shared memory only the instructions need to be parallelized. When the memory is split, it usually requires substantially more programming work that explicitly controls the exchange of information between memory units.

As for a single processor core, a parallel calculation may be limited by 1) the number of arithmetic operations, 2) memory size, 3) the time to move data between memory and processor, and 4) input/output, but communication between processors needs to be considered also. If two processor cores share a calculation, but one of them has to wait for a result from the other, it might take *longer* than on a single processor. Once a new number is written to a register, it needs to make its way up through several levels of caches, to the lowest level of memory shared between the cores (Figure 8.1). If processed by the same core, the number could instead be immediately read from the register, and only the final result needs to end up in main memory. Transposing a large matrix, a simple procedure that requires no floating-point operations but lots of data movement, can be particularly slow on parallel hardware.

Parallel computing is only efficient if communication between the processors is meager. This exchange of information limits the maximum number of processors that can be used economically. The fewer data dependencies, the better the "scalability" of the problem, meaning that the speedup is close to proportional to the number of processors. If the fraction of the runtime that can be parallelized is p, then the execution time on N cores is $t_N = \frac{p}{N}t_1 + (1-p)t_1$, where t_1 is the execution time on a single core. The speedup is

$$S = \frac{t_1}{t_N} = \frac{1}{\frac{p}{N} + 1 - p}$$

This common sense relation is called Amdahl's law. Suppose a subroutine that takes 99% of the runtime on a single processor core has been perfectly parallelized, but not the remaining part of the program, which takes only 1% of the time. Amdahl's law shows that 4 processors provide a speedup of 3.9; the speedup is subproportional to the number of processors. No matter how many processors are used ($N \to \infty$), the speedup is always less than 100, demanded by the 1% fraction of nonparallelized code, $S < 1/(1-p)$. In our example, 100 processors provide a speedup of only 50. For a highly parallel calculation, small non-parallelized parts of the code become a bottleneck. The lesson from Amdahl's law is that the bottleneck will soon lie in the non-parallelized parts of the code.

Amdahl's law considers a problem of fixed size, and determines the speedup achieved from parallel processing. Alternatively, one can also consider how much bigger of a problem can be solved with parallel processing. The answer is

$$S = 1 - p + Np$$

also known as Gustafson's Law. Using, as above, $p = 0.99$ and $N = 100$ a 99-times larger problem can be solved within the same runtime. This distinction is sometimes also referred to as strong versus weak scaling. Strong scaling considers a fixed problem size (Amdahl's law) and weak scaling a fixed problem size per processor (Gustafson's law). Weak scaling is always better than strong scaling, and hence easier to accomplish.

When the very same program needs to run only with different input parameters, the scalability is potentially perfect. No intercommunication between processors is required during the calculation. The input data are sent to each processor at the beginning and the output data are collected at the end. Computational problems of this kind are called "embarrassingly parallel." This form of parallel computing is embarrassingly easy to accomplish, at least comparatively, and doing so is not embarrassing at all; it is an extremely attractive and efficient form of parallel computing.

"Distributed computing" involves a large number of processors located usually at multiple physical locations. It is a form of parallel computing, but the communication cost is very high and the platforms are diverse. Distributed computer systems can be realized, for example, between computers in a computer lab, as a network of workstations on a university campus, or with idle personal computers from around the world. Tremendous cumulative computational power can be achieved with the sheer number of available processors. Judged by the total number of floating-point operations, distributed calculations rank as the largest computations ever performed. In a pioneering development for distributed computing, SETI@home utilizes millions of personal computers to analyze data coming from a radio telescope listening to space.

9.3 PROGRAMMING AND UTILIZING PARALLEL HARDWARE

An effort-free way to take advantage of parallelism on any level is index-free notation, such as arithmetic with an array, e.g., `a(:)=b(:)+1`. This capability is available in some languages, and not in others. In Matlab notation, matrix multiplication `A*B` and element-wise matrix multiplication `A.B` are valid index-free expressions.

Summing numbers can obviously take advantage of parallel processing. But in the sequential implementation

```
s=a[0];
for(i=1;i<N;i++) s=s+a[i];
```

the dependency of s on the previous step spoils parallelization when programmed in this form. The sequential implementation disguises the natural parallelizability of the problem. In Fortran, Matlab, and NumPy, a special command, `s=sum(a)`, makes the parallelizability evident to the compiler, so these intrinsic functions can be expected to be faster than a simple self-made loop.

Generally, an executable we generated runs on a single core, not simultaneously on the multiple cores of a CPU. If we submit a second job, it will be placed by the operating system on a different core and not compete with the first. We can verify this by using the `top` command, which shows the fraction of cycles/resources used by an individual job.

```
> a.out &
> b.out &
> top
  PID   USER     PR NI  ...  S  %CPU %MEM  TIME+  COMMAND
31900   norbert  20  0  ...  R 100.0  0.0  0:50.54 a.out
31901   norbert  20  0  ...  R 100.0  0.0  0:37.20 b.out
```

The first two lines show the simplest way to run two programs in parallel— by submitting each as a background job; the & symbol accomplishes that (see Appendix A). The output of the top command shows a multitude of information, including the process id (pid). The %CPU column shows the percentage of the cycles consumed in *one* core, so both jobs get a whole core. There are other operating-system and user-based processes running also, but they consume only a tiny fraction of the resources or they run on other cores, allowing a utilization of 100% for each of our processes. Once there are more jobs than cores, they compete for CPU cycles, if the jobs are compute-bound.

If we want a single executable to use several cores, we usually have to do something extra. Parallelization of a program can be done by compilers automatically, but if this does not happen in an efficient way—and often it does not—the programmer has to provide instructions, by placing special commands into the source code. This is often done with "compiler directives", an example thereof will be given below.

A common standard for parallelization of programs on shared memory machines is OpenMP (Multi-Processing), and many compilers offer intrinsic support for OpenMP. It specifies, for example, what loop should be parallelized. Table 9.2 shows two code fragments with OpenMP compiler directives. In Fortran, ! indicates a comment line, so the lines starting with !$OMP will be ignored during regular compilation. With OpenMP enabled, the compiler recognizes the lines are compiler directives, rather than an intrinsic part of the language. That way, the exact same source code can be executed with and without OpenMP parallelization by turning a compiler option on and off: `gfortran -fopenmp openmpexample.f90` versus `gfortran openmpexample.f90`. The latter version ignores the OpenMP instructions and runs as a sequential program. In C, the OpenMP commands would begin with `#pragma omp` instead of `!$OMP`.

The result of the program fragment to the left in Table 9.2 is that "Hello world" will be written to the display as many times as the executable thinks it can do so in parallel. This number is set by a system-internal variable (called an "environment variable"), which is typically the number of available

TABLE 9.2 Examples of parallel programming for a shared memory machine using OpenMP compiler directives. The example to the left spawns a task over several threads. The example to the right is a parallelized loop. In Fortran, compiler directives begin with !$OMP.

```
!$OMP PARALLEL              !$OMP PARALLEL DO
print *,'Hello world'       do i=1,10
!$OMP END PARALLEL              ...
```

threads, but the number can also be set by the user. Commonly each core has two threads, so a 4-core processor would output it eight times.

The example on the right performs as many iterations in the loop as possible simultaneously. Its output will consist of ten results, but not necessarily in the same order as the loop.

To write a sequential program and then add directives to undo its sequential nature, is awkward and rooted in a long history when computing hardware was not as parallel as it is now. A more natural solution is to have language intrinsic commands. The index-free notation is one example. If a and b are arrays of the same shape and length, then a=sin(b) can be applied to every element in the array, as it was in the program example of Table 4.2. In Fortran, the right side of Table 9.2 is replaced by do concurrent (i=1:10). But even if a language has such intrinsic commands, it does not guarantee that the workload will be distributed over multiple cores.

For the parallelization of loops and other program blocks, a distinction needs to be made between variables that are shared through main memory (called shared variables) and those that are limited in scope to one thread (called private variables). For example in

```
do concurrent (i=1:10)
   x=i
   do k=1,100000000
      x=sin(3*x)
   enddo
end do
```

it is essential that x and k be private variables; otherwise, these variables could easily be overwritten by another of the ten concurrent tasks, because they may be processed simultaneously. On the other hand, if we chose to use x(i) instead x, the array should be shared; otherwise, there would be ten arrays with ten entries each. If not specified in the program, the compiler will apply default rules to designate variables within a concurrent or parallel block as private or shared, for example, scalar variables as private and arrays as shared.

That much about programming for shared memory parallel machines. A widely accepted standard for parallelization on *distributed* memory machines is MPI (Message Passing Interface). Message passing is communicating between multiple processes by explicitly sending and receiving information.

On a computer cluster, executables are typically submitted through a "job scheduler", such as *Condor, Portable Batch System (PBS)*, and *Slurm*. In about its simplest form, a PBS script may look like this:

```
# request 4 cpu cores
#PBS -l nodes=1:ppn=4
# request 2 hours and 30 minutes of cpu time
#PBS -l walltime=02:30:00
# run the program
/path_to_executable/program_name
exit 0
```

'#' starts a comment, whereas '#PBS' is a command. This script will run the executable on four CPU cores. The walltime is the maximum time the job is allowed to run. The job will be terminated if it exceeds this time, whether or not it is finished, so the walltime better be safely above the actual execution time. It is used to prioritize short jobs over long jobs on a crowded cluster.

Computing clusters also have the organizational unit "node". A node is usually a physical unit such as a board with one or more CPUs and a main memory. Often many of these nodes are stacked into a rack. The hierarchy is the following:

$$\text{cluster or cloud} \supset \text{rack} \supset \text{node} \supseteq \text{CPU} \supset \text{core}$$

On the command prompt, the following submits the job to the cluster. Assuming the file above is called `myscript.pbs`,

```
> qsub myscript.pbs
```

The script is submitted from a master node, and there is no need to ever log into the various compute nodes.

Job schedulers make the use of computer clusters perfectly scalable. We request 1,000 processor cores in the same way we would use a few cores. It also makes sure only one job runs on a core (or thread) at any time. For this reason, a job scheduler can serve a purpose even on a computer with only one CPU and one user. The script also allows for additional controls, such as what disks to write to, or it can email a notification when the job is finished.

A job scheduler can also be used to run a number of independent jobs, the same executable with different input parameters, something known as "job array" or "array job". An environment variable, such as $PBS\_ARRAYID$, distinguishes one job from another, and this integer can be fed to the executable to instruct it to read different input files and create distinguishable output file names.

9.4   HARDWARE ACCELERATION

Arithmetic Co-Processors. CPUs have to carry out versatile tasks, in addition to floating-point arithmetic. Additional dedicated floating-point units can benefit numerically intensive calculations. Arithmetic co-processors, that provide additional floating-point power in support of the CPU, have appeared repeatedly in the history of computer architectures. Currently, co-processors are available in the form of many-core processors, designed for massive parallelism. One example is the Xeon Phi, which has on the order of 64 cores. Even more cores are packed into programmable general-purpose GPUs.

Graphics Processing Units (GPUs), which, as the name indicates, evolved from graphics hardware, provide massively parallel computational capabilities. GPUs consist of many (up to thousands of) cores and each core can execute many threads concurrently. In fact, each core processes threads in groups no smaller than 32 (a "warp").

The many floating-point units are only part of the reason why GPUs achieve high throughput; exploiting memory structures is necessary to keep them busy. Physically, the memory technology of a GPU is qualitatively not much different from that of a CPU; there is fast on-chip memory and slow off-chip memory, but how memory is organized and used is different. For example, part of the memory can be dedicated as read-only memory. And communication among a group of threads is through fast shared memory. The programmer has a more detailed control of memory organization and access patterns.

Programming GPUs efficiently can be difficult, but the payoff is potentially high. GPUs work well only with data-level parallel problems, not with task-parallel problems. And they are only efficient if many FLOPs are carried out for each byte of data moved from main memory to GPU memory. For example, matrix addition is not worthwhile on GPUs, but matrix multiplication is. Successfully ported applications have led to a hundredfold speedup compared to a CPU core (e.g., the gravitational N-body problem). Languages for GPU programming include CUDA and OpenCL. A modern GPU can carry out calculations with single- and double-precision numbers (their ancestors, graphics chips, only require single precision). As we have learned previously, on a CPU, single-precision calculations are no faster than double-precision calculations, but on a GPU the latter are slower.

Problem-customized computer architectures. For particularly large and important problems, specialized hardware optimized for the problem at hand can be considered. Examples of special purpose processors that have been developed and built are GRAPE, for the numerical solution of the gravitational N-body problem, and Deep Blue, for chess playing. So far, such attempts have turned out to be short-lived, because the performance of mainstream processors has increased so rapidly. By the time the specialized system was completed, its performance gain barely competed with the newest mainstream computing hardware. However, since Moore's law has slowed down this approach may become fruitful.

A more mundane approach is to use common hardware in a configuration that is customized for the computational method. If the problem is limited by the time it takes to read large amounts of data, then a set of disks read in parallel will help. If an analysis requires huge amounts of memory, then, well, a machine with more than average memory will help or, if the problem is suitably structured, a distributed memory system can be used. With remote use of computing clusters and computational clouds that occupy not only rooms, but entire floors, and even warehouse-size buildings, a choice in hardware configuration is realistic.

Recommended Reading: Instructional material for OpenMP can be found at `https://computing.llnl.gov/tutorials/openMP/` and elsewhere. For MPI (Message Passing Interface) documentation see `www.mpich.org`. Kirk & Hwu, *Programming Massively Parallel Processors: A Hands-on Approach* is a pioneering book on GPU computing. The aforementioned textbook by Patterson & Hennessy also includes an introduction to GPUs. The following website maintains a list of the fastest supercomputers in the world, which provides an idea of what is computationally feasible: `https://www.top500.org/lists/top500/`.

EXERCISES

9.1 The fragment of a Fortran program below finds the distance between the nearest pair among N points in two dimensions. This implementation is wasteful in computer resources. Can you suggest at least 4 simple changes that should improve its computational performance? (You are not asked to verify that they improve performance.)

```
! ... x and y have been assigned values earlier ...
m=1E38
do i=1,N
   do j=1,N
      if (i==j) cycle  ! skips to next iteration
      r(i,j) = sqrt((x(i)-x(j))**2. + (y(i)-y(j))**2.)
      if (r(i,j)<m) m=r(i,j)
   enddo
enddo
! m is minimum distance
```

9.2 Learn how to ingest a command line argument into a program, e.g., `myprog 7` should read an integer number or single character from the command line. Then use this input argument to read file `in.7` and write a file named `out.7`. This is one approach to run the same program with many different input parameters. Submit the source code.

9.3 Learn how to profile your code, that is, obtain the fraction of time consumed by each function in your program. Take or write an arithmetically intensive program that takes at least 30 seconds to execute, then find out which function or command consumes most of the time. Submit an outline of the steps taken and the output of the profiler.

9.4 In a programming language of your choice, learn how to use multiple cores on your CPU. Some may do this automatically; others will need explicit program and/our launch instructions. Verify that the program runs on multiple cores simultaneously.

## CHAPTER 10

The Operation Count; Numerical Linear Algebra

10.1 INTRODUCTION

Many computations are limited simply by the sheer number of required additions, multiplications, or function evaluations. If floating-point operations are the dominant cost then the computation time is proportional to the number of mathematical operations. Therefore, we should practice counting. For example, $a_0 + a_1 x + a_2 x^2$ involves two additions and three multiplications, because the square also requires a multiplication, but the equivalent formula $a_0 + (a_1 + a_2 x)x$ involves only two multiplications and two additions.

More generally, $a_N x^N + ... + a_1 x + a_0$ involves N additions and $N + (N - 1) + ... + 1 = N(N + 1)/2$ multiplications, but $(a_N x + ... + a_1)x + a_0$ only N multiplications and N additions. The first takes about $N^2/2$ FLOPs for large N, the latter $2N$ for large N. (Although the second form of polynomial evaluation is superior to the first in terms of the number of floating-point operations, in terms of roundoff it may be the other way round.)

A precise definition of the "order of" symbol O is in order (big-O notation). A function is of order x^p if there is a constant c such that the absolute value of the function is no larger than cx^p for sufficiently large x. For example, $2N^2 + 4N - \log(N) + 7$ is $O(N^2)$. Although the number may be larger than $2N^2$, it is, for sufficiently large N, always smaller than, say, $3N^2$, and therefore $O(N^2)$. More generally, a function is of order $g(x)$ if $|f(x)| \leq c|g(x)|$ for $x > x_0$. For example, $\log(N^2)$ is $O(\log N)$, because it is never larger than $2\log N$. The analogous definition is also applicable for small numbers, as in chapter 6, except the inequality has to be valid for sufficiently small instead of sufficiently large values. The leading order of the operation count is the "asymptotic order count". For example, $2N^2 + 4N - \log(N) + 7$ asymptotically takes $2N^2$ steps.

With the definition $f(N) \leq cN^p$ for $N > N_0$, a function f that is $O(N^6)$ is also $O(N^7)$, but it is usually implied that the power is the lowest possible. Some call a "tight" bound big-Θ, but we will stick to big-O. A tight bound means there are constants $k > 0$ and K such that $kN \leq f(N) \leq KN$ for sufficiently large N. Or, in more general form, a function f is of order $g(x)$ if there are constants $k > 0$ and K such that $k|g(x)| \leq |f(x)| \leq K|g(x)|$ for $x > x_0$. (It can be argued that $f(x) \in O(g(x))$ is a more precise notation than $f(x) = O(g(x))$, because it is a "one-way equality"; we would not write $O(g(x)) = f(x)$. But the equality sign is commonly accepted and can be viewed as corresponding to the verb "is".)

After this discourse, back to actual counting. Multiplying two $N \times N$ matrices, $c_{ij} = \sum_{k=1}^{N} a_{ik}b_{kj}$, obviously requires N multiplications and $N - 1$ additions for each output element. Since there are N^2 elements in the matrix this yields a total of $N^2(2N - 1)$ floating-point operations, or about $2N^3$ for large N, that is, $O(N^3)$.

An actual comparison of the relative speed of floating-point operations was given in Table 8.1. According to that table, we do not need to distinguish between addition, subtraction, and multiplication, but divisions take longer.

The expression $1/\sqrt{x^2 + y^2}$ requires one addition, two multiplications, one division, and a square root. The square root is equivalent to about 10 FLOPs. The division is a single floating-point operation, but it takes about four times longer than the faster FLOPs. If we strictly counted the number of FLOPs, a division would count as one, but because we count FLOPs for the purpose of evaluating computational speed, it should count for more. In units of the fast and basic FLOPs $(+, -, \times)$, evaluation of this expression takes the equivalent of about 3+10+4 such FLOPs.

10.2   OPERATION COUNTS IN LINEAR ALGEBRA

Slow and fast method for the determinant of a matrix

It is easy to exceed the computational ability of even the most powerful computer. Hence methods are needed that solve a problem quickly. As a demonstration we calculate the determinant of a matrix. Doing these calculations *by hand* gives us a feel for the problem.

One way to evaluate a determinant is Cramer's rule, according to which the determinant can be calculated in terms of the determinants of submatrices. Although we are ultimately interested in the $N \times N$ case, a 3×3 matrix can serve as an example. Cramer's rule using the first row is

$$\det \begin{pmatrix} a_{11} & a_{12} & a_{13} \\ a_{21} & a_{22} & a_{23} \\ a_{31} & a_{32} & a_{33} \end{pmatrix} =$$

$$= a_{11} \det \begin{pmatrix} a_{22} & a_{23} \\ a_{32} & a_{33} \end{pmatrix} - a_{12} \det \begin{pmatrix} a_{21} & a_{23} \\ a_{31} & a_{33} \end{pmatrix} + a_{13} \det \begin{pmatrix} a_{21} & a_{22} \\ a_{31} & a_{32} \end{pmatrix}$$

And for a 2×2 matrix the determinant is

$$\det \begin{pmatrix} a_{11} & a_{12} \\ a_{21} & a_{22} \end{pmatrix} = a_{11} \det (a_{22}) - a_{12} \det (a_{21}) = a_{11}a_{22} - a_{12}a_{21}$$

For example, for the matrix

$$A = \begin{pmatrix} 1 & 1 & 0 \\ 2 & 1 & -1 \\ -1 & 2 & 5 \end{pmatrix}$$

this yields $\det(A) = 1 \times (5 + 2) - 1 \times (10 - 1) + 0 \times (4 + 1) = 7 - 9 = -2$.

For a matrix of size N this requires calculating N subdeterminants, each of which in turn requires $N - 1$ subdeterminants, and so on. More precisely, the cost evolves as $T_N = NT_{N-1} + 2N - 1$. Once at size 2, $T_2 = 3$, and at size 1, $T_1 = 0$. Hence the number of necessary operations is at least $N!$. That surely increases rapidly with N. For example, for $N = 100$, $T_N \approx 10^{158}$. Even with 10^9 floating-point operations per second and all of the world's computers, we could not never finish such a calculation.

A faster way of evaluating the determinant of a large matrix is to bring the matrix to upper triangular or lower triangular form by linear transformations. Appropriate linear transformations preserve the value of the determinant. The determinant is then the product of diagonal elements, as is clear from the previous definition; in upper triangular form and choosing the last row to apply Cramer's rule, only one subdeterminant matters. For our example, A, the transforms $(\text{row}2 - 2 \times \text{row}1) \to \text{row}2$ and $(\text{row}3 + \text{row}1) \to \text{row}3$ yield zeros in the first column below the first matrix element. Then the transform $(\text{row}3 + 3 \times \text{row}2) \to \text{row}3$ yields zeros below the second element on the diagonal:

$$\begin{pmatrix} 1 & 1 & 0 \\ 2 & 1 & -1 \\ -1 & 2 & 5 \end{pmatrix} \to \begin{pmatrix} 1 & 1 & 0 \\ 0 & -1 & -1 \\ 0 & 3 & 5 \end{pmatrix} \to \begin{pmatrix} 1 & 1 & 0 \\ 0 & -1 & -1 \\ 0 & 0 & 2 \end{pmatrix}$$

Now, the matrix is in triangular form and $\det(A) = 1 \times (-1) \times 2 = -2$.

An $N \times N$ matrix requires N such steps; each linear transformation involves adding a multiple of one row to another row, that is, N or fewer additions and N or fewer multiplications. Hence this is an $O(N^3)$ procedure. Below we will show that the leading order is $(2/3)N^3$. Therefore calculation of the determinant by bringing the matrix to upper triangular form is far more efficient than the previous method using Cramer's rule. For say $N = 10$, the change from $N!$ to N^3 means a speedup of over several thousand. This enormous speedup is accomplished through a better choice of numerical method.

Non-intuitive operation counts in linear algebra

We all know how to solve a linear system of equations by hand, by extracting one variable at a time and repeatedly substituting it in all remaining equations, a method called Gauss elimination. This is essentially the same as we have done above in consecutively eliminating columns. The following symbolizes the procedure again on a 4×4 matrix:

$$\begin{pmatrix} **** \\ **** \\ **** \\ **** \end{pmatrix} \rightarrow \begin{pmatrix} **** \\ *** \\ *** \\ *** \end{pmatrix} \rightarrow \begin{pmatrix} **** \\ *** \\ ** \\ ** \end{pmatrix} \rightarrow \begin{pmatrix} **** \\ *** \\ ** \\ * \end{pmatrix}$$

Stars represent nonzero elements and blank elements are zero. For an $N \times N$ matrix, eliminating the first column takes about $2N^2$ floating-point operations, the second column $2(N-1)^2$, the third column $2(N-2)^2$, and so on. This yields a total of about $2N^3/3$ floating-point operations. (One way to see that is to approximate the sum by an integral, and the integral of N^2 is $N^3/3$.)

Once triangular form is reached, the value of one variable is known and can be substituted in all other equations, and so on. These substitutions require only $O(N^2)$ operations. A count of $2N^3/3$ is less than the approximately $2N^3$ operations for matrix multiplication. Solving a linear system is faster than multiplying two matrices.

During Gauss elimination the right-hand side of a system of linear equations, $Ax = b$, is transformed along with the matrix. Many right-hand sides can be transformed simultaneously, but they need to be known in advance.

Inversion of a square matrix can be achieved by solving a system with N different right-hand sides. The first column of $AA^{-1} = I$, where I is the identity matrix, is

$$A \begin{pmatrix} c_{11} \\ c_{21} \\ c_{31} \end{pmatrix} = \begin{pmatrix} 1 \\ 0 \\ 0 \end{pmatrix}$$

where c_{ij} are elements of the inverse matrix A^{-1}. Since the right-hand side(s) can be carried along in the transformation process, this is still $O(N^3)$, and not more. Given $Ax = b$, the solution $x = A^{-1}b$ can be obtained by multiplying the inverse of A with b, but it is *not* necessary to invert a matrix to solve a linear system. Solving a linear system is faster than inverting and inverting a matrix is faster than multiplying two matrices.

We have only considered computational efficiency for these matrix operations. It is also desirable to avoid dividing by too small a number, optimize roundoff behavior, or introduce parallel efficiency. Since the solution of linear systems is an important and ubiquitous application, all these aspects have received exhaustive attention and elaborate routines are available. *BLAS* (Basic Linear Algebra Subprograms) and *LAPACK* (Linear Algebra Package) are legendary and widely available numerical libraries for linear algebra.

Figure 10.1 shows the actual time of execution for a program that solves a linear system of N equations in N variables. First of all, note the tremendous computational power of computers: solving a linear system in 1000 variables, requiring about 700 million floating-point operations, takes less than one second. And it was performed on a single CPU core. In chapter 8 it was claimed that one second is enough to solve a linear system with thousands of variables. This demonstrates that this is indeed the case. The increase in computation

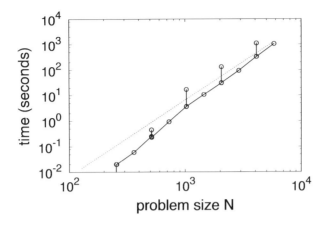

FIGURE 10.1 Execution time of a program solving a system of N linear equations in N variables. For comparison, the dashed line shows an increase proportional to N^3.

time with the number of variables is not always as ideal as expected from the operation count, because time is required not only for arithmetic operations but also for data movement. For this particular implementation the execution time is larger when N is a power of two. Other wiggles in the graph arise because the execution time is not exactly the same every time the program is run.

10.3   OPERATION COUNTS FOR A FEW COMMON METHODS

Operation count also goes under the name of "computational cost", "computational complexity", or "time complexity" (of an algorithm). Table 10.1 shows the operation count for several important problems, or, more precisely, of common methods for these problems, as there can be more than one method for each.

As demonstrated above, solving linear systems, matrix inversion, and matrix multiplication are all $O(N^3)$. A triangular matrix is of the form shown in Figure 10.2. A tridiagonal system of equations can be solved with $O(N)$ effort, not merely $O(N^2)$ but $O(N)$, as compared to $O(N^3)$ for the full matrix. If there are elements in the upper-right and lower-left corner, it is called a "cyclic tridiagonal matrix", which can still be solved in $O(N)$ steps.

A matrix where most elements are zero is called "sparse", and a tridiagonal matrix is an example of a sparse matrix. The opposite of sparse is a dense or full matrix. A sparse matrix can be solved, with appropriate algorithms, much faster and with less storage than a full matrix.

Sorting N floating-point numbers takes $O(N \log N)$ steps. One of many

TABLE 10.1 Asymptotic order counts for several important problems solved by standard algorithms.

problem	operation count	
Solution of N linear equations in N variables	$2N^3/3$	$O(N^3)$
Inversion of $N \times N$ matrix	N^3	$O(N^3)$
Matrix multiplication	$2N^3$	$O(N^3)$
Inversion of tridiagonal $N \times N$ matrix	$45N$	$O(N)$
Sorting N real numbers	$N \log_2 N$	$O(N \log N)$
Fourier Transform of N equidistant points	$5N \log_2 N$	$O(N \log N)$

$$
\begin{pmatrix}
* & * & & & & \\
* & * & * & & & \\
 & * & * & * & & \\
 & & \ddots & \ddots & \ddots & \\
 & & & * & * & * \\
 & & & & * & *
\end{pmatrix}
$$

FIGURE 10.2 A tridiagonal matrix. Blank elements are zero.

methods that accomplishes that will be described in chapter 12. Per element this is only a $O(\log N)$ effort, so sorting is "fast".

A Fourier transform is calculated as,

$$
\hat{F}_k = \sum_{j=0}^{N-1} f_j e^{-2\pi ikj/N}
$$

with $k = 1, ..., N$. Every Fourier coefficient $\hat{F}_k$ is a sum over N terms, so the formula suggests that a Fourier transform of N points requires $O(N^2)$ operations. However, already over two hundred years ago it was realized that it can be accomplished in only $O(N \log N)$ steps. This classic algorithm is the Fast Fourier Transform (FFT). The FFT is a method for calculating the discrete Fourier transform. Specifically, the operation count is about $5N \log_2 N$. The exact prefactor does not matter, because for a non-local equation such as this, the time for data transfer also has to be considered.

For the FFT algorithm to work, the points need to be uniformly spaced both in real as well as in frequency space. The original FFT algorithm worked with a domain size that is a power of two, but it was then generalized to other factors, combinations of factors, and even prime numbers. A domain of any size can be FFT-ed quickly, even without having to enlarge it to the next highest power of two (and worry what to fill in for the extra function values).

Methods can be grouped into complexity classes. $O(1)$ is called constant-time complexity. Algorithms that take $O(N)$ steps are called "linear (time) algorithms", and those that take $O(N \log N)$ steps "quasi-linear algorithms", because they are effectively almost as fast as $O(N)$ algorithms. An algorithm is said to be solvable in polynomial time, if the number of required steps is $O(N^k)$ for some nonnegative integer k. Polynomial-time algorithms could also be called power-law algorithms, but somehow this does not have the same ring to it. Exponential complexity refers to $2^{O(N)} = e^{O(N)}$. The "space complexity" of an algorithm describes how much memory the algorithm needs in order to operate.

Obviously, $O(2^N)$ is not the same as $O(e^N)$, because there is no N-independent prefactor for 2^N that would make it larger than e^N. However, with the order-of in the exponent, these are equivalent for any base larger than one. To see this

$$e^{O(N)} \le e^{cN} = (e^c)^N = \left(\frac{e}{2}\right)^c 2^{cN} = 2^{cN + c \log_2(e/2)} = 2^{O(N)}$$

This only proves an inequality, $e^{O(N)} \le 2^{O(N)}$, but the reverse inequality also holds, $2^{O(N)} \le e^{O(N)}$. If an inequality is true in both directions, it must be an equality.

Brainteaser: Show that $\log(N)$ is never large. Hence, linear, $O(N)$, and quasi-linear, $O(N \log N)$, algorithms are practically equally fast. While the mathematical function $\log(N)$ undoubtedly always increases with N and has no upper bound, for practical purposes, its value is never large. Consider situations such as the largest counter in a single loop and the number of distinguishable floating-point numbers in the interval 1 to 10.

10.4 DATA MOVEMENT AND DATA LOCALITY

Calculations may not be limited by the number of arithmetic operations per second but by data movement; hence we may also want to keep track of and minimize access to main memory. The ratio of floating-point operations to bytes of main memory accessed is called the "arithmetic intensity." It has units of FLOPs/byte. Large problems with high arithmetic intensity are floating-point limited, while large problems with low arithmetic intensity are limited by memory bandwidth. Another way of thinking about the arithmetic intensity is as FLOPs/second divided by bytes/second (bandwidth), which again has units of FLOPs/byte.

A matrix transpose has an arithmetic intensity of zero. Vector addition of double-precision numbers requires 1 FLOP for each output element, and three numbers of memory traffic, two from main memory to the CPU and one in the opposite direction, so the arithmetic intensity is $1/(3 \times 8)$ FLOPs/byte, very low. Computing the iteration $x_{n+1} = 4x_n(1 - x_n)$ for $n = 1, ..., N$ requires $3N$ FLOPs and transfer of only the initial and final value, so the arithmetic intensity is $3N/8$ FLOPs/byte. This is clearly a floating-point limited task.

For multiplication of two $N \times N$ matrices, simple row times column multiplication involves $2N$ memory pulls for each of the N^2 output elements. If the matrix elements are 8-byte double-precision numbers, then the arithmetic intensity is $(2N-1)/(8 \times (2N+1)) \approx 1/8$ FLOPs/byte. At such a low arithmetic intensity the bottleneck is likely the memory bandwidth.

At current technology, a performance of several GFLOPS per core is common and transfer rates of GB/s are common, so the cross-over arithmetic intensity is at a few FLOPs/byte. Of course, an application may achieve neither the maximum possible FLOPS nor the maximum available bandwidth, if it is not efficiently programmed.

Data traffic from main memory can sometimes be reduced by *tiling*. For example, a single C loop `for(i=0; i<N; i++)` can be tiled by replacing it with

```
for(j=0; j<N; j+=B)
  for(i=j; i<min(N,j+B); i++)
```

where B is the size of the block (tile).

Tiled matrix multiplication. The idea behind this method is to divide the matrix into tiles (or blocks) of size $B \times B$, small enough that they fit into a fast (on-chip) memory, such as the lowest level cache (Figure 10.3). For each pair of input tiles, there are $2B^2$ movements from the (slow) main memory to the processor. All arithmetic is carried out on them, and then the tiles are replaced with the next pair of input tiles. The partial sums are kept only in local memory. For simplicity, we assume N is a multiple of B; then we have to go through N/B pairs of tiles. The output matrix consists of $(N/B)^2$ tiles. In total, this amounts to $(N/B)^3 2B^2 = 2N^3/B$ inward memory movements. At the end, N^2 elements are moved to main memory, but because $B \ll N$ this does not change the leading term. Hence, there are $2N/B$ data movements for each matrix element, compared to $2N$ for the naive method. The number of floating-point operations is still the same, so the arithmetic intensity has changed from $1/8$ to $B/8$. The size of B is limited by the physical size of fast memory. This reduces data access to the main memory and turns matrix multiplication into a truly floating-point limited operation.

$$\begin{pmatrix} \vdots & \vdots & \vdots \\ \vdots & B & \vdots \\ \vdots & \vdots & \vdots \end{pmatrix} = \begin{pmatrix} \vdots & \vdots & \vdots \\ B & B & B \\ \vdots & \vdots & \vdots \end{pmatrix} \begin{pmatrix} \vdots & B & \vdots \\ \vdots & B & \vdots \\ \vdots & B & \vdots \end{pmatrix}$$

FIGURE 10.3 Two matrices are multiplied using square tiles to improve data locality. The tiles involved in the calculation of one of the output tiles are labeled.

There are enough highly optimized matrix multiplication routines around that we never have to write one on our own, but the same concept—grouping of data to reduce traffic from and to main memory—can be successfully applied to various problems with low arithmetic intensity.

Recommended Reading: Golub & Van Loan, *Matrix Computations* is a standard work on numerical linear algebra. A chapter in the aforementioned book by Patterson & Hennessey deals with arithmetic intensity.

EXERCISES

10.1 How many times is the content of the innermost loop executed?

```
for (i=0; i<N; i++) {
  for (j=i+1; i<N; j++) {
    for (k=j+1; k<N; k++) {
      ...
    }
  }
}
```

(If unfamiliar with the syntax of a C loop, see chapter 4.)

10.2 Order the following functions from lowest to highest. Consider O as a tight bound. If any are of the same order, indicate which:
N, 3^N, $N \log N$, $2N - N^3 + 3N^5$, $N + \log N$, N^2, $N!$, $(2N)!$, $(\log N)^2$, $\sqrt{N} + \log N$, $\log_2 N$, $\ln N$, $\log(5N^2 + 7)$, $\log(\log N)$, 7

10.3 Show that

a. $(2N)!/(2^N N!) = O(N!)$
b. $\log(N!) = O(N \log N)$
c. $2^{O(\log N)} = O(N^k)$

10.4 Calculate the arithmetic intensity for the 3-dimensional gravitational N-body problem when evaluated with a simple nested double loop.

a. Write down program code or pseudocode that calculates the acceleration of each body in 3D.
b. Count the number of floating-point operations required for the evaluation, to leading order. Assume a square root operation is equivalent to 10 FLOPs.
c. Count the number of bytes that need to be accessed from main memory. Assume floating-point numbers have 8 bytes.
d. Take the ratio. Based on this ratio, do you think this calculation is floating-point limited or data-transfer limited?

## CHAPTER 11

Random Numbers & Stochastic Methods

11.1 GENERATION OF PROBABILISTIC DISTRIBUTIONS

Random number generators are not truly random, but use deterministic rules to generate "pseudorandom" numbers, for example with

$$x_{n+1} = 16808\, x_n \bmod (2^{31} - 1)$$

that is, the remainder of $16808x_i/2147483647$. The starting value x_0 is called the "seed." Pseudorandom number generates of the form $x_{n+1} = (ax_{n+1} + b)$ mod m, where all numbers are integers, are known as "linear congruential generators".

Pseudorandom number generators can never ideally satisfy all desired statistical properties. For example, the above type of generator repeats after at most m numbers. More generally, with only finitely many computer representable numbers, the sequence will ultimately always be periodic, though the period can be extremely long. Random number generators can be faulty. Particular choices of the parameters or seed can lead to short periods. And the coefficients in formulae like the one above need to be chosen carefully. Some implementations of pseudorandom number generators were deficient, but a good implementation suffices for almost any practical purpose. If there is any doubt about the language-intrinsic random number generator, then vetted open-source code routines can be used. (The general code repositories listed in Appendix B are one place to the find them.)

There have been attempts to use truly random physical processes, such as radioactive decay, to generate random numbers, but technical limitations implied that the sequences were not perfect either. Moreover, an advantage of deterministic generators is that the numbers are reproducible as long as one chooses the same seed. (And if we never want them to start the same way, the time from the computer's clock can be used for the seed.)

Pseudorandom number generators produce a uniform distribution of numbers in an interval, typically either integers or real numbers in the interval from 0 to 1 (without perhaps one or both of the endpoints). How do we obtain a different distribution? A new probability distribution, $p(x)$, can be related to a given one, $q(y)$, by a transformation $y = y(x)$. The probability to be between x and $x + dx$ is $p(x)dx$. By construction, this equals the probability to be between y and $y + dy$. Hence,

$$|p(x)dx| = |q(y)dy|$$

where the absolute values are needed because y could decrease with x, while probabilities are always positive. If $q(y)$ is uniformly distributed between 0 and 1, then

$$p(x) = \begin{cases} |dy/dx| & \text{for } 0 < y < 1 \\ 0 & \text{otherwise} \end{cases}$$

Integration with respect to x and inverting yields the desired transformation. For example, an exponential distribution $p(x \geq 0) = \exp(-x)$ requires $y(x) = \int_0^x p(x')dx' = -\exp(-x) + 1$ and therefore $x(y) = -\ln(1 - y)$. This equation transforms uniformly distributed numbers into exponentially distributed numbers. The distribution has the proper bounds $x(0) = 0$ and $x(1) = +\infty$. In general, it is necessary to invert the integral of the desired distribution function $p(x)$. That can be computationally expensive, particularly when the inverse cannot be obtained analytically.

Alternatively the desired distribution $p(x)$ can be enforced by rejecting numbers with a probability $1 - p(x)$, using a second randomly generated number. These two methods are called "transformation method" and "rejection method," respectively.

That said, many common distributions, in one or more variables, are available from built-in functions, and if not, the reference at the end of the chapter may contain a prescription on how to generate them.

11.2 MONTE CARLO INTEGRATION: ACCURACY THROUGH RANDOMNESS

Besides obvious uses of random numbers there are numerical methods that intrinsically rely on probabilistic means. A representative example is the Monte Carlo algorithm for multi-dimensional integration.

Consider the following method for one-dimensional integration. We choose random coordinates x and y, evaluate the function at x, and determine whether y is below or above the graph of the function; see Figure 11.1. If this is repeated with many more points over a region, then the fraction of points that fall below the graph is an estimate for the area under the graph relative to the area of the entire region. This is Monte Carlo integration.

Intuitively this seems to be a crude method for integration, because the function value is only used to decide whether it is larger or smaller than

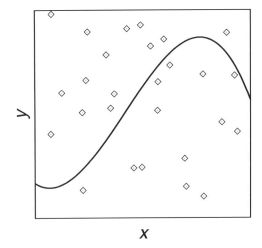

FIGURE 11.1 Randomly distributed points are used to estimate the area below the graph.

the random value, but otherwise discarded. Let's quantify the error of this numerical scheme.

Suppose we choose N randomly distributed points, requiring N function evaluations. How fast does the integration error decrease with N? The probability of a random point to be below the graph is proportional to the area a under the graph. Without loss of generality, the constant of proportionality can be set to one. The probability of having m specific points below the graph and $N - m$ specific points above the graph is $a^m(1 - a)^{N-m}$, in the sense that the first point is below, the second above, the third below, and so on, whereas the same result would be obtained if the third was above and the second below. The points are interchangeable, so there is a binomial factor. The probability P of having any m points below the graph and any $N - m$ points above the graph is

$$P(m) = \binom{N}{m} a^m (1 - a)^{N-m}$$

An error E can be defined as the root mean square difference between the exact area a and the estimated area m/N:

$$E^2 = \sum_{m=0}^{N} \left(\frac{m}{N} - a\right)^2 P(m)$$

This sum is $E^2 = (1 - a)a/N$. When the integral is estimated from N sample points, the error E is proportional to $1/\sqrt{N}$. For integration in two or more rather than one variable, the exact same calculation applies.

A conventional summation technique to evaluate the integral has an error too, due to discretization. With a step size of h, the error would be typically $O(h^2)$ or $O(h^4)$, depending on the integration scheme and what is known at the integration boundaries (Chapter 6). In one dimension, h is proportional to $1/N$, so the rate of convergence is $O(1/N^2)$ or $O(1/N^4)$ and it makes no sense to use Monte Carlo integration instead of conventional numerical integration techniques.

In more than one variable, conventional integration requires more function evaluations to achieve a sufficient resolution in all directions. For N function evaluations and d variables the grid spacing h is proportional to $N^{-1/d}$. Generously assuming the error is $O(h^4)$, the rate of convergence will be $N^{-4/d}$. Hence, in many dimensions, the error decreases extremely slowly with the number of function evaluations. With, say, ten billion, 10^{10} function evaluations for an integral over ten variables, there will only be 10 grid points along each axis. One can speak of a "dimension barrier," because the number of function evaluations required for a certain accuracy increases exponentially with the number of dimensions. The accuracy of Monte Carlo integration, on the other hand, is the same for any number of integration variables, $O(N^{-1/2})$. It is more efficient to distribute the ten billion points randomly and measure the integral in this way. Another way to put it: It's not that Monte Carlo integration is fast; it's that multi-dimensional integrals are so slow.

Integration is undoubtedly a deterministic procedure. The Monte Carlo method illustrates not only a method of integration, but also the following: Statistical methods can be used efficiently to solve deterministic problems! Surely, statistical methods are appropriate for statistical problems, but what was shown here is that sometimes statistical methods can be miraculously efficient for deterministic problems.

Another advantage of Monte Carlo integration that should not go unmentioned is that irregularly shaped boundaries can be easily accommodated. For a many-dimensional deterministic integral, imposing complex boundary conditions can be a headache. Also, it is possible to improve upon the $O(N^{-1/2})$ convergence rate if one manages to restrict the domain closer to the function.

11.3 SAMPLE PROBLEM: ISING MODEL*

It is overdue that we again go through an example motivated by a specific scientific question. The problem comes from statistical mechanics, and it is relevant both to physics and chemistry. Moreover, it is one of those relatively simple numerical calculations that can solve problems that are famously difficult to treat mathematically.

The Ising model consists of a regular lattice where each site has a "spin" which points up or down. The spins are thought of as magnets that interact with each other. The energy E at each site is in this model determined by the nearest neighbors (n.n.) only: $E_i = -J\sum_{(\text{n.n.})} s_i s_j$, where the spin s is $+1$ or -1, and J is a positive constant. The lattice can be in one, two, or more dimensions. In one dimension there are two nearest neighbors, on a two-dimensional square lattice four nearest neighbors, and so on. There is no real physical system that behaves exactly this way, but it is a simple model to study the thermodynamics of an interacting system. Ferromagnetism is the closest physical analog, and the sum of all spins corresponds to magnetization. (Like magnetic poles repel each other, and the energy is lowest when neighboring magnets have opposite orientations. Hence, it would appear we should choose $J < 0$ in our model. However, electrons in metals interact in several ways and in ferromagnetic materials the energies sum up to align electron spins. For this reason we consider $J > 0$.)

The spins have the tendency to align with each other to minimize energy, but this is counteracted by thermal fluctuations. At zero temperature all spins will align in the same orientation to reach minimum energy (either all up or all down, depending on the initial state). At nonzero temperatures will there be relatively few spins opposite to the overall orientation of spins, or will there be a roughly equal number of up and down spins? In the former case there is macroscopic magnetization; in the latter case the average magnetization vanishes.

According to the laws of statistical mechanics the probability to occupy a state with energy E is proportional to $\exp(-E/kT)$, where k is the Boltzmann constant and T the temperature. This is also known as the "Boltzmann factor". In equilibrium the number of transitions from up to down equals the number of transitions from down to up. Let $W(+ \to -)$ denote the probability for a flip from spin up to spin down. In steady state the probability P of an individual spin to be in one state is proportional to the number of sites in that state, and hence the equilibrium condition translates into

$$P(+)W(+ \to -) = P(-)W(- \to +)$$

For a simulation to reproduce the correct thermodynamic behavior, we hence need

$$\frac{W(+ \to -)}{W(- \to +)} = \frac{P(-)}{P(+)} = \exp\left[-\frac{E(-) - E(+)}{kT}\right]$$

Here, $E(-)$ is the energy when the spin is down, which depends on the orientation of the neighbors, and likewise for $E(+)$. For the Ising model the ratio on the right-hand side is $\exp(2bJ/kT)$, where b is an integer that depends on the orientations of the nearest neighbors.

There is more than one possibility to choose the transition probabilities $W(+ \to -)$ and $W(- \to +)$ to achieve the required ratio. Any of them will lead to the same equilibrium properties. Denote the energy difference between before and after a spin flip with ΔE, defined to be positive when the energy increases. One possible choice is to flip from the lower-energy state to the higher-energy state with probability $\exp(-\Delta E/kT)$ and to flip from the higher-energy state to the lower-energy state with probability one. If $\Delta E = 0$, when there are equally many neighbors pointing in the up and down direction, then the transition probability is taken to be one, because this is the limit for both of the preceding two rules. This method, which transitions to a higher energy with probability $\exp(-\Delta E/kT)$ and falls to a lower energy with probability 1, is known as the "Metropolis algorithm."

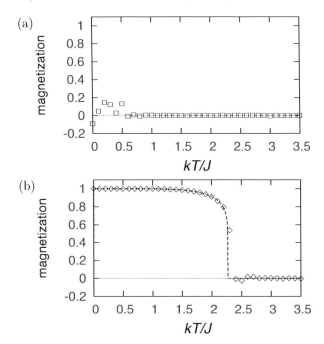

FIGURE 11.2 Magnetization versus temperature from simulations of the Ising model in (a) one dimension (squares) and (b) two dimensions (diamonds). The dashed line shows the analytical solution for an infinitely large two-dimensional system, $[1 - 1/\sinh^4(2J/kT)]^{1/8}$.

A simulation of the Ising model with the Metropolis algorithm requires only a short program, which uses a random number generator. For example, in one dimension and in C, the core of the program would contain

```
if (E>=0.) s[i]=-s[i];
if (E<0.) {  // initially in low energy state
  x=ran1(&idum);
  if (x<exp(E/T)) s[i]=-s[i];
}
```

E has previously been calculated and is $+2$, 0, or -2, and `ran1` is an externally defined function that provides uniformly distributed numbers from 0 to 1, so the probability for $x < C$ to occur is C. For a two- or three-dimensional grid these program lines would be the same, only s would have more indices. For $T = 0$ the exponent diverges, but the computer may even handle that correctly by evaluating the exponential as zero. If we were into parallel computing, these lines would be (in Fortran)

```
where (E>=0.) s=-s
x=random(idum)
where (E<0. .and. x<exp(E/T)) s=-s
```

although that generates more random numbers than needed.

Although the program is short and simple, validation is, as always, prudent. For a one-dimensional chain of spins, we can simply look at the sequence of spin changes for a segment. The program produces the following flips for the one-dimensional model: $+ - + \rightarrow + + +$, $+ + - \rightarrow + - -$, and either $+ + + \rightarrow + + +$ or $+ + + \rightarrow + - +$. Considering the middle one of the three spins, it starts out with a high energy state and therefore must flip. The second transition starts with zero energy state, and the spin must therefore also flip. And $+ + +$ is a low-energy state, and therefore has some probability to flip or not to flip. In conclusion, the program's output behaves exactly as intended.

Figure 11.2 shows the magnetization as a function of temperature obtained with such a program. Part (a) is for the one-dimensional Ising model and the spins are initialized in random orientations. The scatter of points at low temperatures arises from insufficient equilibration and averaging times. In one dimension the magnetization vanishes for any temperature, with some ambiguity at very low T. (It turns out that the magnetization vanishes for any temperature larger than zero.)

But in two dimensions there are two phases. Part (b) of Figure 11.2 shows the magnetization for the two-dimensional Ising model, where initially all spins point up. At low temperatures there is magnetization, but at high temperatures the magnetization vanishes. The two phases are separated by a continuous, not a discontinuous, change in magnetization. As the system size increases this transition becomes more and more sharply defined. For an infinite system, the magnetization vanishes beyond a specific temperature $T_c \approx 2.269 J/k$, the "critical temperature."

Figure 11.3 shows snapshots of the spin configuration in the two-dimensional Ising model, at temperatures below, close to, and above the phase transition. At low temperatures, panel (a), most spins are aligned in the same direction, with a few exceptions. Occasionally there are individual spins that flip. In this regime, spins correlate over long distances, although the interactions include only nearest neighbors. At higher temperatures there are more fluctuations and more spins of the opposite orientation. Above the phase transition, panel (c), there are a roughly equal number of spins up and down, clustered in small groups. The fluctuations dominate and there is no macroscopic magnetization.

We now set out to find a theoretical explanation for this behavior. For the Ising model the total energy of the system can change with time. Let $\Omega(E)$ be the number of spin configurations with total energy E, where E is the sum

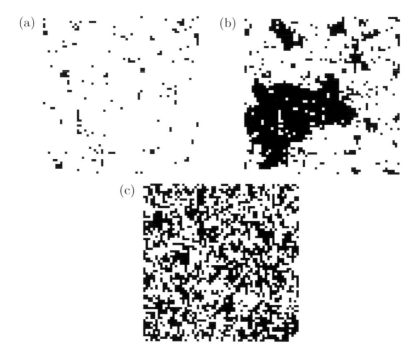

FIGURE 11.3 Snapshot of spin configurations for the Ising model (a) below, (b) close to, and (c) above the critical temperature. Black indicates positive spins, white negative spins.

of the energies of all individual spins, $E = \sum_j E_j$. The probability to find the system in energy E is proportional to

$$\Omega(E) \prod_j e^{-E_j/kT} = \Omega(E)e^{-\sum_j E_j/kT} = \Omega(E)e^{-E/kT}$$

With $F = E - kT \ln \Omega(E)$ this expression can also be written as $\exp(-F/kT)$. The quantity $S = k \ln \Omega(E)$ is also known as "entropy", and F is the so-called "Free Energy". For a given temperature, the system is thus most often in a configuration that minimizes F, because it makes the exponential largest. The quantity F becomes smaller when the energy E is lowered or when the entropy S increases. Low energy configurations prevail at low temperature and form the ordered phase of the system; high entropy configurations prevail at high temperature and form the disordered phase.

The thermodynamic properties of the Ising model can be obtained analytically if one manages to explicitly count the number of possible configurations Ω as a function energy E. (To be more precise, not only does one need to count the number of possible configurations for each energy, but also determine the magnetization of these configurations.) This can be done in one dimension, but in two dimensions it is much, much, much harder. The dashed line in Figure 11.3(b) shows this exact solution, first obtained by Lars Onsager. In three dimensions no one has achieved it, and hence we believe that it is impossible to do so. The underlying reason why the magnetization can be obtained with many fewer steps with the Metropolis algorithm than by counting states, is that a probabilistic method automatically explores the most likely configurations. The historical significance of the Ising model stems largely from its *analytical* solution. Hence, our numerical attempts in one and two dimensions have had a merely illustrative nature. But even simple variations of the model (e.g., the Ising model in three dimensions or extending the interaction beyond nearest neighbors) are not solved analytically. In these cases numerics is valuable and no more difficult than the simulations we have gone through here.

Recommended Reading: For generation and testing of random numbers see Knuth, *The Art of Computer Programming, Vol. 2: Seminumerical Algorithms*. Methods for generating probability distributions are found in Devroye, *Non-Uniform Random Variate Generation*, which is also available on the web at http://luc.devroye.org/rnbookindex.html.

EXERCISES

11.1   Uniform distribution of points on sphere. For points that are distributed uniformly over the surface of a sphere (the same number of points per area), the geographic coordinates are not uniformly distributed. Here we seek to generate point coordinates that are uniformly distributed over a unit sphere.

    a.   Derive the transformation necessary to calculate geographic latitude λ and longitude ϕ from random variables u, v that are uniformly distributed between 0 to 1.

    b.   Implement a program that produces random points uniformly distributed over a sphere.

    c.   Devise a way to validate the outcome.

    d.   Validate your implementation with this test.

11.2   Generate a Maxwell distribution for velocity $|v| = \sqrt{v_1^2 + v_2^2 + v_3^2}$ by generating Gaussian distributions for each of the three velocity components with standard deviation 1. Verify that the standard deviation of the resulting Maxwell distribution is indeed what it should be. Include a convergence test that demonstrates that the standard deviation of $|v|$ and of each component v_i approaches the theoretical values as the number of randomly generated points increases.

11.3 A probability distribution of the form

$$p(x) = \frac{1}{\pi} \frac{a}{a^2 + x^2}$$

is known as Cauchy or Lorentzian distribution.

a. Use the transformation method to generate a Cauchy distribution from a uniform probability distribution.
b. Conduct a convergence test for the standard deviation, and show that it behaves as it should.
c. To verify that the generated probability distribution is indeed of the correct shape, use a "Kolmogorov-Smirnov" test. It uses the maximum difference between the cumulative distribution of the ideal and the sample distribution.

$$D_N = \max_x |\hat{P}_N(x) - P(x)|$$

and compares it to a threshold value for that sample size. The cumulative distribution is defined by $P(x) = \int_{-\infty}^{x} p(x')dx'$, and similarly for the sample. The threshold is not all that easy to calculate itself, but for $N > 35$ and a significance level of 0.1, $D = 1.22/\sqrt{N}$ is a reasonable approximation.

## CHAPTER 12

Algorithms, Data Structures, and Complexity

12.1   AN EXAMPLE ALGORITHM: HEAPSORT

Heapsort is a sorting algorithm for floating-point numbers which is fast even in the worst case scenario. It exploits data arrangement and demonstrates that a good algorithm does not have to be straightforward. Hence, several interesting concepts can be learned from it at once.

Suppose N real numbers need to be ordered. The algorithm works with triplets of numbers, and arranges them in a "heap" as shown in Figure 12.1. In the example one floating-point number is included among the integers, because if all numbers were integers there would be a faster way to sort them. That one floating-point number serves as a reminder that this is a sorting method for real numbers not merely integers.

We start by taking three numbers and select the largest among them. Then we take the next number and compare it with one of the numbers at the foot of the previous triplet, arranging the data in a heap as illustrated in Figure 12.1(a). If the upper number in any triplet is not the largest, it is swapped with the larger of the two numbers beneath it. "Any triplet" includes triplets at all levels, so more than one swap may be necessary when a large number is added at the bottom of the heap. At the end, the largest element is on top. The final arrangement of data after the first stage of the algorithm is shown in the rightmost tree of Figure 12.1(a).

The second stage of the algorithm starts with the largest element, on top, and replaces it with the largest element on the level below, which is in turn replaced with its largest element on the level below, and so on. In this way the largest element is pulled off first, then the second largest, third largest,

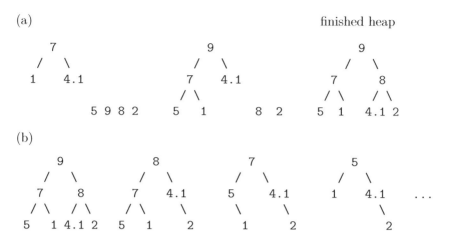

FIGURE 12.1 Example of heapsort algorithm applied to the unsorted sequence 7, 1, 4.1, 5, 9, 8, 2. (a) construction of the heap; (b) extraction of the maximum. The final result is 9, 8, 7, 5, ...

and so on, and all numbers are eventually sorted according to their size; see Figure 12.1(b).

Denote the smallest integer power of 2 larger than N with N'. When $N = 1$, $N' = 2$; when N is 2 or 3, then $N' = 4$, and so on. The number of levels in the heap is $\log_2 N'$. The first stage of the algorithm, building the heap, loops through the N elements, but the potential swaps at each element addition means we might have to go through each level, so the first stage requires up to $O(N \log N') = O(N \log N)$ work. In the second stage, comparing and swapping is necessary up to N times for each level of the tree. Hence each of the two stages of the algorithm is $O(N \log N)$, and the total is also $O(N \log N)$. Considering that merely going through N numbers is $O(N)$ and that $\log N$ is usually a small number, sorting is "fast."

A binary tree as in Figure 12.2 can simply be stored as a one-dimensional array. The first level occupies the first element, the two elements of the next level the following two elements, and so on. With b levels, the length of this array is $2^b - 1$. The index in the array a_j for the i-th element in the b-th level of the tree is $2^{b-1} + i - 1$, where b, i, and j all start at 1.

Heapsort is also highly memory efficient, because elements can be replaced in-place. It is never necessary to create a copy of the entire list of elements.

Brainteaser: Sort one suit of a deck of cards with the heapsort algorithm.

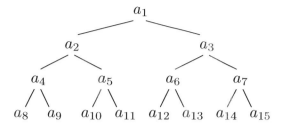

FIGURE 12.2 A binary tree indexed as one-dimensional array with elements $a_1, ..., a_{15}$.

12.2 DATA STRUCTURES

Trees, which we have encountered in the heapsort algorithm, are a "data structure". The difference between a (full) tree and a heap is that the right-left order matters in a tree but not in a heap. Trees come with a beautifully self-explanatory terminology: master node, nodes, parents, children, branches, and leaves. These terms mean what we guess they mean. A tree with two children per node is called a binary tree, and with four children it is a called a quadtree.

Arrays are another, simple data structure. Its elements are stored sequentially in memory. (In C there is a direct relation between array indices and memory addresses, and incrementing one is as good as incrementing the other.)

A further possibility is to store pointers to data, that is, every data entry includes a reference to where the next entry is stored. Such a storage arrangement is called "list." Inserting an element in a sequence of data is faster when the data are stored as a list rather than as an array. An element can be quickly added to the end of an array, but inserting an element elsewhere in an array requires all subsequent elements to be shifted. Insert is an $O(1)$ operation for a list and an $O(N)$ operation for an array. On the other hand, accessing the last element is faster in an array than in a list. Accessing an element is $O(N)$ for a list and $O(1)$ for an array. Lists cause cache misses (described in chapter 8), because sequential elements are often not stored sequentially in memory, in contrast to the cache's locality assumption.

Table 12.1 shows a classification of common data structures. Data structures that are indivisible or "atomic" are called "primitive". These are most often built-in data types, and we commonly know these as "data types" rather than "data structures". Data structures that are not atomic are called non-primitive, composite, or compound. These are often user-defined, but some languages, such as C++ and Java, support many types of data structures. (In contrast, low-level languages, such as assembly or machine code, have no data types.) Compound structures can further be divided into linear and nonlinear structures. In a linear data structure the elements are organized sequentially. Examples are arrays, simple lists, stacks (last in, first out), and queues (first

TABLE 12.1 Classification of data structures

Primitive Data Structures:		boolean/logical
		integer
		float/double
		complex
		character
		pointer
Compound Data Structures:	Linear:	array
		(linked) list
		stack
		queue
	Nonlinear:	tree
		graph

in, first out). Even when their indices are sequential, it does not imply adjacent elements are stored sequentially in memory. As already mentioned, arrays use consecutive memory addresses; lists generally do not. In a nonlinear data structure the elements can be connected to more than one other element. Examples are trees and graphs (networks).

Another useful classification of data structures is in terms of their lifecyle during program execution. "Static data structures" are those whose size and associated memory location are fixed at compile time. "Dynamic data structures" can expand or shrink during program execution and the associated memory location can move. Allocation of an array during runtime (in C with `malloc`) is one example of a dynamic data type.

Hashing is used to speed up searching. Consider the problem of searching an array of length N. If the array is not sorted, this requires $O(N)$ steps. If the array is sorted, a divide-and-conquer approach can be used, which reduces the number of required steps to $O(\log N)$ (Exercise 12.2). The search would be even faster if there was a function which would point to the sought value in the array. A "hash function" is a function, which when given a value, generates an address in a table. Understandably, such a mathematical function that would provide a one-to-one relation between values and indices can often not be constructed, but even an imperfect relation, where two or more different values may hash to the same array index, is helpful. Values that collide can be put in a linked list. A hash table then is an array of lists. Exercise 12.1 will go through one example. Hash functions can be constructed empirically, and even applied to character strings by converting the strings to ASCII code. They enable efficient lookup: given a string (e.g., the name of a chemical compound), find corresponding entries(e.g., the chemical formula and other associated information).

Concurrent data structure. Accessing data with a single process is one thing, sharing the same data among concurrent processes is another. By definition, the order at which concurrent processes will execute is not known, so if they share the content of a variable there can be conflicts. Consider a simple global counter c=c+1. If one of the concurrent processes needs to increment the counter, it has to make sure another process is not also incrementing it simultaneously. Even on a shared memory machine, registers, the level-1 cache, or even higher levels of cache, are local to the CPU core (Figure 8.1), so the content of the variable would have to transverse these levels to make sure the value of the counter is correct, and during this time the process stalls. Here is a way to avoid write conflicts for a (grow-only) counter: Each process has its own local counter c[i], and when the value of the global counter is needed, all the local counters are added. There will be a query function that calculates the sum, C = sum(c[i]), and a function that updates the local counter, c[i]=c[i]+1. This is an example of a "conflict-free" data type, because it avoids write conflicts. Often it is not practical to use standard data structures for concurrent processing, and special data structures are needed to store and organize the data.

12.3 COMPUTATIONAL COMPLEXITY & INTRACTABLE PROBLEMS

It is rarely possible to prove it takes *at least* a certain number of steps to solve a problem, no matter what algorithm one can come up with. That raises the question: "Isn't there a faster way to do this?" A famous example is multiplication of two $N \times N$ matrices in $O(N^{\log_2 7})$ steps, $\log_2 7 = 2.8...$, which is less than $O(N^3)$. However, the prefactor of the operation count for this asymptotically faster algorithm is impractically large. The operation count of the fastest possible algorithm is called the "computational complexity" of the problem (as opposed to the computational complexity of the *algorithm*).

The number of necessary steps can increase very rapidly with problem size. Not only like a power, such as N^3, but as $N!$ or $\exp(N)$. These are computationally unfeasible problems, because even for moderate N they cannot be solved on any existing computer, e.g., $100! \approx 10^{158}$; even 10^9 FLOPs a second can barely make a difference. A year has only $86400 \times 365 \approx 3.1 \times 10^7$ s, conveniently memorized as $\pi \times 10^7$ s. A single CPU core may provide up to 10^{17} FLOPs per year. Even with thousands or millions of cores, we get nowhere close to 10^{158}. A problem is called computationally "intractable" when the required number of steps to solve it increases faster than any power of N. For example, combinatorial problems can be intractable when it is necessary to try more or less all possible combinations. Complexity classes that are exponential or worse are referred to as non-polynomial or "NP".

Examples of intractable problems are integer factorization, the Traveling Salesman Problem, and the longest common subsequence. They will be briefly described here, without proof that they require non-polynomial time algorithms. Integer factorization is about quickly finding the factors that make up a given integer. If it is a prime number, it has no such factors. If it is the product of two large prime numbers, the factorization becomes algorithmically so challenging that it is used for cryptography. No existing computer can factor the number, but it is easy to verify whether a factor works, so owning one of the factors serves as "key." The Traveling Salesman Problem asks: Given a list of cities and the distances (or airfares) between each pair of cities, what is the shortest possible route that visits each city and returns to the origin city? Another example of an intractable problem is finding the longest common subsequence among several strings. A subsequence is the string with elements deleted. For example, ATG is a subsequence of CATAGC. The elements have the same order but do not have to be contiguous. This problem arises in spell checking and genome comparisons. (Genomes contain sequences of nucleotide bases and these bases are abbreviated with single letters. DNA consists of only four types of bases, abbreviated as A, C, G, and T.) Finding the longest common subsequence (Figure 12.3) requires an algorithm which is exponential in the number of sequences. So if we wish to know the longest consecutive sequence of bases among a large number of genomes, it is a tough problem. The field of "bioinformatics" deals with problems such as this.

AGTGGACTTTGACAGA

AGTGGACTTAGATTTA

TGGATCTTGACAGATT

AGTTGACTTACGTGCA

ATCGATCTATTCACCG

FIGURE 12.3 Five sequences consisting of letters A, C, G, and T. The longest common subsequence is TGACTTAG.

For some problems it has been proven that it is impossible to solve them in polynomial time; others are merely believed to be intractable since nobody has found a way to solve them in polynomial time. (Incidentally, proving that finding the longest common subsequence, or any of a large number of equivalent problems, cannot be accomplished in polynomial time is one of the major outstanding questions in present-day mathematics. Nobody has yet proved that this group of problems indeed requires an algorithm with "NP"-complexity rather than only polynomial or "P"-complexity.)

A small problem modification can sometimes turn a hard problem into an easy one. (A humorous response, with a grain of truth, to the Traveling Salesman Problem is: selling online involves only $O(0)$ travel.) A computer scientist or mathematician may know more about algorithms than a scientist, but the scientist better understands the context of the problem and can make approximations or modifications to the problem formulation that elude those who look at the problem from a purely formal perspective. For this reason, scientists have a role to play in research on algorithmically challenging problems.

Another venue of approach to intractable problems are algorithms that find the solution only with a probability, a very high probability. In this context, it is sobering to remember that at any time a cosmic ray particle can hit the CPU or main memory and flip a bit and cause an error. This is rare, but does occur. In an airplane, which is higher up and exposed to more cosmic and solar radiation, a crucial computation has to be carried out on duplicate processors (or threefold actually, because if two processors get different answers, it is ambiguous which one is correct). On spacecraft, which are exposed to even more radiation, radiation-caused bit flips are routinely observed. Even for earthly applications, memory technology often incorporates "error-correcting code" (ECC). An extra bit is used that stores whether the number of 1s in a binary sequence is odd or even; in other words the "parity" of a binary sequence. If one bit flips, the parity changes, which reveals the error. If two bits flip, the parity does not change and the error remains undetected. But the probability of an event to occur twice is the square of the probability for it to occur once, and therefore extremely small. For a scientific computation, we do not worry about absurdly unlikely sources of error. In other words, a probabilistic algorithm that errs with a probability of, say, 10^{-100} is as reliable as a deterministic algorithm.

12.4   APPROXIMATIONS CAN REDUCE COMPLEXITY

Finite precision can reduce complexity

The integral of certain functions cannot be obtained in closed form. For example, the integral of $\exp(-x^2)$ cannot be expressed in terms of elementary functions, or at least not by a finite number of elementary functions. Yet, the integral is simply the Error function, a special function. Are special functions more difficult to evaluate numerically than elementary functions? Not profoundly. Either type of function can be expanded in a series that converges fast or approximated by some kind of interpolating function. A computer evaluates an elementary function, such as sine, the same way. It uses an approximation algorithm, a rapidly converging series for instance. In the end, the hardware only adds, subtracts, multiplies, and divides. There is no fundamental definition of "elementary function"; it is merely a convention. Elementary functions are the ones commonly available on calculators and within programming languages. They are most likely to have ready-made, highly efficient implementations. Calculating an elementary function, a special function, or a function with no name at all to 16-digit precision is fundamentally the same kind of problem. For example, the Error function can be calculated to six digits of precision with only 23 floating-point operations, with the following expression (from Abramowitz & Stegun, *Handbook of Mathematical Functions*):

$$\mathrm{erf}(x) \;=\; \frac{2}{\sqrt{\pi}} \int_0^x e^{-t^2}\,dt$$

$$\approx 1 - \frac{1}{(1 + a_1 x + a_2 x^2 + a_3 x^3 + a_4 x^4 + a_5 x^5 + a_6 x^6)^{16}}$$

where $a_1 = 0.0705230784$, $a_2 = 0.0422820123$, $a_3 = 0.0092705272$, $a_4 = 0.0001520143$, $a_5 = 0.0002765672$, and $a_6 = 0.0000430638$. For $x < 0$ use $-\mathrm{erf}(|x|)$. (Rearranging the polynomial can reduce the number of floating-point operations from 23 to 18, but may worsen roundoff.)

Finite precision can substantially reduce the computational complexity of a problem. Many programming languages provide an `erf` function the same way they provide the `sin` function. What matters is not whether there is a closed-form expression, that is, a way to evaluate a function with a finite number of terms, but whether an efficient approximation is available. In the end, the result will be a number with finite precision.

This brings us to a philosophical or fundamental point about computation. The expression $\sqrt{2}$ is considered an "exact" answer. The digits of an irrational number never repeat, so to actually obtain all its digits requires infinitely many steps, plus the number is infinitely long. A more tangible definition of an "exact" result is to argue it can be calculated to any desired precision within a finite number of steps. With that definition, any method that can produce a result to a desired precision after a finite number of steps is equally "exact". One could also argue that numerical results are limited by roundoff/truncation error. The registers of a CPU have fixed byte length, which ultimately causes this truncation. But if needed, arbitrary precision calculations can be emulated on a software level. That will be much slower computationally, but in principle truncation errors can be made arbitrary small for a numerical calculation. The only difference is that $\sqrt{2}$ is written with fewer symbols than, say, the Newton iteration of chapter 2, and that is hardly of fundamental significance. A way to view the world is from the perspective of "computability"; what matters fundamentally is whether and how a result can be computed to a desired precision.

A case in point are the roots of high-degree polynomials. Long ago, mathematician Évariste Galois proved (not long before he died in a duel) that only polynomials up to 4th degree have roots that can always be expressed with closed-form expressions. Polynomials of 5th and higher degree also have solutions, but in general these cannot be expressed in closed form. Numerically, roots of high-degree polynomials can be easily obtained to arbitrary precision. A twist to this story are 4th degree polynomials, whose roots can be expressed with closed-form expressions, but these expressions are generally so large that they are of limited usefulness.

An approximation that reduces complexity: Hierarchical N-body method*

When an exact equation cannot be fully evaluated, because it takes too many steps, one way or the other we will have to settle for an approximate result. For problems that are computationally limited, the relevant question becomes how to minimize the error for a given computational resource. One such example was already described in chapter 11: for a fixed number of function evaluations, the probabilistic error of Monte-Carlo integration is smaller than the deterministic error of another method. Here we revisit the N-body pairwise interaction problem, and show how systematically stopping a calculation as soon as it reaches a certain error requirement can dramatically reduce the operation count.

The formula for the gravitational force of N objects on each other,

$$\mathbf{F}_i = Gm_i \sum_{j=1,j\neq i}^{N} m_j \frac{\mathbf{r}_i - \mathbf{r}_j}{|\mathbf{r}_i - \mathbf{r}_j|^3}$$

with $i = 1, ..., N$, inevitably involves $O(N^2)$ operations. For example, the gravitational interaction of a 100 billion stars in a galaxy cannot be simulated with direct evaluation of all pairwise forces. Even if taking into account that i and j can be swapped, the expression in the sum needs to be evaluated about $N^2/2$ times. With three spatial components, $\mathbf{r} = (x, y, z)$, one element of the sum is

$$m_j \frac{x_i - x_j}{\sqrt{(x_i - x_j)^2 + (y_i - y_j)^2 + (z_i - z_j)^2}^3}$$

This amounts to about $3 \times 3 + 8 = 17$ FLOPs per term, plus the square root, which costs the equivalent of roughly 10 FLOPs (Table 8.1). So, in total the cost is around 30 FLOPs per element in the sum. With $N = 10^{11}$ that amounts to roughly $30 \times 10^{22}/2 \approx 10^{23}$ FLOPs. And the forces need to be re-evaluated at every time step.

If, however, the result is calculated only to a limited precision, nearby objects can be grouped together to obtain the force they exert on all objects far away from them. This leads to a dramatic reduction in the number of pairwise forces that need to be evaluated.

What is the error made by replacing the gravitational pull of a group of objects with that of their center of mass? To find out we need to expand $\mathbf{r}/|\mathbf{r}|^3$, or the potential $1/|\mathbf{r}|$, around a location other than zero. The result is called a multi-pole expansion (an expansion in inverse powers of $|\mathbf{r}|$), but this would be too lengthy of a digression here. We will simply assume that a group of masses can be replaced with a single mass, if (in a two-dimensional setting) their extent is within a threshold angle Θ.

The Barnes-Hut algorithm is a hierarchical tree-based algorithm for the gravitational N-body problem. It uses an adaptive grid that successively subdivides the domain until there is only one particle in each box (Figure 12.4a). The boxes are subdivided as particles are added. To identify which box a particle belongs to, its coordinates can be multiplied by a factor and rounded to an integer. An interactive tool that visualizes the hierarchical grid building can be found, e.g., at www.stefanom.org/wgc/test_tree.html?v=0.

A convenient data structure for the adaptive grid is an (incomplete) quadtree in two dimensions or an octree in three dimensions (Figure 12.4b). The left-to-right order of the children matters and is chosen clockwise in the figure. The particles are at the leaves of the tree. Each node corresponds to

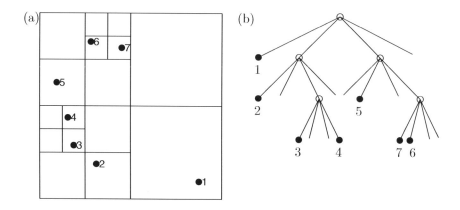

FIGURE 12.4 (a) Adaptive grid for the N-body problem in two dimensions and (b) the corresponding quadtree. Tree nodes right to left correspond to the NE, NW, SW, and SE quadrants.

a box and contains the total mass and center of mass coordinates for all the particles it contains.

If the ratio D/r = (size of box) / (distance from particle to center of mass of box) is small enough, then the gravitational force due to all the particles in the box is approximated by their center of mass. To evaluate the forces, we start at the head of the tree and calculate the distance, and move down the tree, until the criterion $D/r < \Theta$ is satisfied, where Θ is a user-supplied threshold. An interactive tool that visualizes the force-evaluation stage of the algorithm can be found, e.g., at www.khanacademy.org/computer-programming/quadtree-hut-tree/1179074380.

Now we will roughly count how many steps this algorithm requires. If the particles are uniformly distributed, each body will end up at the same level in the tree, and the depth of the tree is $\log_4 N$. The cost of adding a particle is proportional to the distance from the root to the leaf in which the particle resides. Hence, the complexity of inserting all the particles (building the tree) is $O(N \log N)$.

The operation count for the force evaluation can be determined by plotting all the boxes around an individual particle that satisfy the geometric criterion $D/r < \Theta$. Their size doubles with distance, and since there are at most three undivided squares at each level (the fourth square being occupied by the particle under consideration), the total amount of work is proportional to the depth in the tree at which the particle is stored. Hence, this cost is also proportional to $O(\log N)$ for each particle. In total, the tree-based N-body method is an $O(N \log N)$ algorithm, at least when the bodies are distributed homogeneously.

In conclusion, systematic truncation or grouping reduces the computational cost of the gravitational N-body problem from $O(N^2)$ to $O(N \log N)$.

For large N, the choice we are faced with is not between an approximate and an exact answer (that only has roundoff), but between an approximate result for N bodies and an exact result for a lot fewer than N bodies. When 100 billion stars are replaced with 1 billion stars each with a hundred times the mass, some dynamical aspects, such as how many stars are ejected, may be quite different. The physics is better represented with a fast approximate evaluation than with a precise evaluation limited to a small sample.

Recommended Reading: Cormen, Leiserson, Rivest, & Stein, *Introduction to Algorithms* is a widely used textbook. Another is Sedgewick & Wayne, *Algorithms*, with the accompanying site `http://algs4.cs.princeton.edu`. For a treatment of the fundamentals of computation see Blum, Cucker, Shub, and Smale, *Complexity and Real Computation*.

EXERCISES

12.1 *Hash Table.* A commonly used method for hashing positive integers is modular hashing: choose the array size M to be prime and for an integer value k, compute the remainder when dividing k by M. This is effective in dispersing the values evenly between 0 and $M - 1$.

Consider the hashing function (value)%7.
a. Generate the indices for the following values: 13, 96, 16, 3, 11, 112, 23, 54, 42.
b. To look up one of these 9 values, how many steps are necessary on average? Count calculation of the remainder as one step, and following a linked list counts as a second step.

12.2 Given a sequence of N real numbers sorted by size. Show that a search for a number in this array takes at most $O(\log N)$ steps.

12.3 The operation count T of a divide-and-conquer method is given by the recursive relation
$$T(N) = 2T(N/2) + O(N)$$
with $T(1) = O(1)$. Find an explicit expression for $T(N)$. Assume N is a power of 2.

## CHAPTER 13

Data

In the modern world of science, data abound. An overview of basic notations and conventions serves as a helpful preliminary for this chapter. The smallest unit anything is stored in on a medium is 1 byte (8 bits). Upper case B stands for byte, and lower case b for bit.

Historically the prefix kilo often means $2^{10} = 1024$ instead of 1000, and mega was used for $2^{20} = 1048576$ instead of 10^6. Use of the same prefixes with two slightly different meanings is confusing. This ambiguity was addressed with the introduction of prefixes that specifically refer to powers of 1024 instead of powers of 1000. The old terms, kilobyte (kB), Megabyte (MB), and Gigabyte (GB), are now supposed to be used for decimal powers only. The new terms are kibibyte (KiB or KB), mebibyte (MiB), and gibibyte (GiB). Table 13.1 lists the prefixes for large decimal and binary numbers. Again it is helpful to remember that $10^3 \approx 2^{10}$.

TABLE 13.1 The classic decimal and the new binary unit prefixes for large numbers.

Decimal			Binary		
kilo	k	$1000^1 = 10^3$	kibi	Ki (K)	$1024^1 = 2^{10}$
mega	M	$1000^2 = 10^6$	mebi	Mi	$1024^2 = 2^{20}$
giga	G	$1000^3 = 10^9$	gibi	Gi	$1024^3 = 2^{30}$
tera	T	$1000^4 = 10^{12}$	tebi	Ti	$1024^4 = 2^{40}$
peta	P	$1000^5 = 10^{15}$	pebi	Pi	$1024^5 = 2^{50}$
exa	E	$1000^6 = 10^{18}$	exbi	Ei	$1024^6 = 2^{60}$
zetta	Z	$1000^7 = 10^{21}$	zebi	Zi	$1024^7 = 2^{70}$
yotta	Y	$1000^8 = 10^{24}$	yobi	Yi	$1024^8 = 2^{80}$

13.1   DATA FILES AND FORMATS

Plain text and binary formats

For readability it is advantageous to create human readable output, that is, plain text. Plain text is also a highly portable file format; except, the end of line encoding can vary between operating systems, which can be a nuisance, until one discovers one of the tools that convert them automatically.

The simplest and most widely accepted form of plain text is ASCII. In ASCII encoding (an "encoding" maps bit patterns to characters), every character is represented by a number between 32 and 127. They include the alphabet of the English language, numerals, and common symbols. For example, the character "A" is 65. Lower case "a" is $97 = 65 + 32$, so the upper-case and lower-case characters differ by only one bit, because 32 is a power of 2. The number "1" has ASCII code 49. This ASCII mapping is universally accepted. ASCII makes use of only 7 bits, so the 8th bit in a byte is wasted. Also, the lowest 32 characters are unprintable. The encoding wastes a bit of space, but that is only a minor loss. Each character is represented by one byte that is understood by basically anything. Hence, ASCII is extremely portable, except for the end of line character.

Characters that are used to represent line breaks are Line Feed (\n) and Carriage Return (\r). They are part of the ASCII set too; their codes are 10 and 13, respectively. The line-ending on Unix, Linux, and Mac operating systems is \n; that for Windows & DOS is \r\n. (The use of two special characters for a newline traces back to mechanical typewriters, where at the end of a line one had to do two things to start the next line: return the carriage to the start position and advance by a line.) Other conventions, \r and \n\r, have also appeared in computer history, but have essentially died out. Nowadays it is either \n or \r\n for a newline.

ASCII is English language centered. Needless to say, 7 bits are far too few to represent all characters in all alphabets of human languages. This limitation is overcome with the so-called UTF-8 character encoding. Here the 8th bit is used to indicate whether the character is stored in one byte or more than one byte. It is a variable-length encoding that reduces perfectly to ASCII for the first 127 characters. At the same time, by using two, three, or more bytes an unlimited number of characters can be encoded. (UTF stands for Unicode Transformation Format, but is more succinctly referred to as Unicode.)

Plain text, such as ASCII and Unicode, is to be contrasted with *binary* format. Binary numbers take advantage of all bits in a byte, and may be organized in units of more than one byte. And there needs to be a key of how to interpret them. The disadvantage of binary files is that we have to worry about how to read them. Often that is no problem, but when the key to read the binary is lost, it can be a fatal problem. Sometimes the metadata to a binary file are in plain text, so simply viewing the file may reveal the required information on how to read the content.

Binary files are portable, except for a flip in endian-ness on different machines; some computers write groups of bytes left-to-right and others right-to-left (big-endian or little-endian). Endianness refers to the ordering of bytes within a multi-byte value. In the big endian convention, the smallest address has the most significant byte (MSB), and in the little endian convention, the smallest address has the least significant byte (LSB). For a single byte representation, as ASCII, there is no ambiguity. When a multi-byte representation is written and read on the same computer, it also works fine. But if binary data are written on one machine and then read on another with the opposite endianness, the output will make no sense at all. UTF-8 stores the endianness of the byte order in itself, so the there is no ambiguity.

File size and compression

It is possible to at least estimate file size. When data are stored in text format, as they often are, each character takes up one byte. Delimiters and blank spaces also count as characters. The number 1.23456E-04, without leading or trailing blanks, takes up 11 bytes on the storage medium. If an invisible carriage return is at the end of the number, then it consumes an additional byte. A four-byte floating-point number has typically six significant digits, such as 1.23456E-04. As a rule of thumb, a set of stored data takes up more disk space than the same set in memory. The analogous argument obviously applies for 8-byte floating-point numbers as well.

A large file containing mostly numbers uses only a small part of the full character set and can hence be substantially compressed into a file of smaller size. Number-only files typically compress, with conventional utilities, to around 40% of their original size. If repetitive patterns are present in the file, the compression will be even stronger.

The ~40% are empirically determined by using numerical output from scientific calculations or random numbers and compressing them with standard utilities. The theoretical reason behind this compression factor can be explained too. One byte represents 256 possibilities but there are only ten numerals, although, it will also be necessary to include spaces, periods, $\pm$ signs, or end of line characters. Assume about 13 different characters appear frequently in a large file. $13 \times 13 = 169$, so two characters easily fit into one byte, making for a compression slightly better than a factor of 2. (Viewed differently, 256 are 8 bits and 13 are about 3.7 bits, which makes for a ratio of $8/3.7 \approx 2.2 \approx 1/0.46$.) Most files have lots of redundancy, so better compression ratios than that can be expected.

Although plain text files can always be compressed into a smaller binary file, no compression method can compress every binary file. The proof is by contradiction. If it could, then compressing the compressed file would yield an even smaller file, until we eventually arrive at a file of size zero.

A compressed file saves storage space, and a smaller file is also faster to read and write. It is faster to read a compressed file and decompress it in memory on the fly than it is to read the uncompressed file.

Data interchange formats

XML (EXtensible Markup Language) was designed to describe data. (HTML was designed to display data.) This format has widespread support for creation, reading, and decoding. Here is a brief example of XML formatted data:

```
<molecule>
    <formula>H2O</formula>
    <name>water</name>
    <weight>18</weight>
</molecule>
```

Those familiar with HTML are well familiar with the syntax `<tag>` to begin a block and `</tag>` to end it. But in XML there are no pre-defined tags, and the author creates whatever tags she desires. One cannot miss to repeat the well known phrase: "XML does not *do* anything."

An alternative to XML is JSON (JavaScript Object Notation). Although originally derived from the JavaScript language, JSON is a language-independent data format. JSON syntax is shorter, so it has lower overhead than XML. For example,

```
{
    "formula": "H2O",
    "name": "water",
    "weight": 18
}
```

Code for parsing and generating XML or JSON data is readily available in many programming languages.

Many data formats are based on XML, and use pre-defined tags (called XML Schemas). For example, for geographic information KML (Keyhole Markup Language) is such a format:

```
<?xml version="1.0" encoding="UTF-8"?>
<kml xmlns="http://www.opengis.net/kml/2.2">
<Document>
<Placemark>
  <name>Mauna Kea</name>
  <description>summit of Hawaii Island</description>
  <Point>
    <coordinates>-155.468066,19.820664,4207</coordinates>
  </Point>
</Placemark>
</Document>
</kml>
```

The difference to (plain) XML is that these tags have specific meanings. (The kml tag on the second line is only for documentation; it does not have any function beyond that.) If this file is fed into a common geographical software (e.g., Google Earth or software that comes with handheld GPS devices), it will know how to interpret the coordinates.

Data that are tagged with XML, JSON, or other systematic formatting are "structured data", as opposed to "unstructured data". They are also examples of "hierarchical data", because tags are nested within tags. Most data are unstructured. Rigorously tabulated and documented data also count as structured data.

Image formats

There are two types of digital image formats: raster graphics and vector graphics. Raster graphics is pixel based and simply stores the value for each pixel; it is used by many common formats: jpg or jpeg, tiff, png, jpg2000, and others. Uncompressed raster images tend be very large. There is *lossy* and *lossless* compression. Classic jpeg is a lossy format, whereas png is lossless, and jpg2000 offers both options. The danger with lossy formats is that repeated file manipulations will lead to an ever increasing degradation of the image quality. On the other hand, even a tiny loss of information might allow for a much smaller file size.

For vector graphics, lines and points are encoded, which means when enlarging a figure, its elements remain perfectly sharp. It also means individual components can be edited. Moreover, text within the graphics is searchable because it can be stored as letters. Common vector graphics file formats are ps (postscript), pdf (portable document format), and svg (scalable vector graphics). (Incidentally, svg is an XML-based data format.) These formats can also store information pixelwise, but that is not what they are meant for. For example, a pdf created by scanning pages of text from paper, without optical character recognition, is not searchable. Vector graphics is intended for line art and formatted text. A diagram or a plot with points and lines for a publication is best stored in a vector graphics format. A photograph of a newly discovered insect species is best stored as raster graphics.

Some image formats can store several layers, e.g., tiff and svg can, whereas png and jpg cannot. Almost all common image formats can carry metadata, for example, the settings of the camera which took the photo or the creator of a graphics.

ImageMagick is free software that performs many standard image processing tasks, such as converting between formats or resizing images. *GraphicsMagick* is a fork of ImageMagick that emphasizes computational efficiency.

An example of a special-purpose data format is GeoTIFF to store maps. GeoTIFF is a public domain metadata standard which allows georeferencing information to be embedded within a TIFF file. This information can include geographic coordinates, the type of map projection, coordinate systems, reference ellipsoid, and more. Any software that can open TIFF images can also open GeoTIFF images, and simply ignore the metadata, while geospatial software can make use of the metadata.

13.2   TEXT PROCESSING UTILITIES

'Text' in 'text processing tools' refers to data in plain text format. Data, whether from laboratory measurements, observations, or from computer simulations, come in different formats, are created on different operating systems, have different symbols for comment lines, and different placeholders for invalid entries. Text processing utilities can search files, selectively print part of the file, replace parts of the file, perform arithmetic operations, and more. Sophisticated automated manipulations can be done quickly with text processing languages such as awk, Perl, sed, and others. These belong to group III in Figure 4.1, and are so useful that they belong to the standard toolbox of scientists.

We start with a utility called "grep". Its name comes from Globally search a Regular Expression and Print, but the connotation "grab" describes it just as well. (Another aspect for the origin of the term is that g/re/p is a valid command in the ed line editor.) The following displays all lines that contain "NaN":

```
grep NaN yourfile
```

This command-line utility is an excellent data filter. grep can search for a string, or the absence of a string: `grep -v NaN yourfile` displays all lines that do not contain "NaN." The option `-i` makes the search case-insensitive. By adding quotes around the string, spaces can be placed in the grep search. And the following

```
grep -f file yourfile
```

reads search strings from a file named 'file', with one search term per line.

Next on our tour is a group of small utilities that are part of the GNU Core Utilities (CoreUtils). `sort` sorts lines of text files; `head` and `tail` output the beginning or end portion of a file, respectively. `paste` merges lines horizontally; it takes two files as input. `wc` prints the number of bytes, words, and lines in files.

`diff` outputs the difference between two files. For example,

```
diff -b old.py new.py
```

outputs the difference between the two files, ignoring any difference in blank spaces (that is what the `-b` option does). No output indicates they are identical.

`tr` ("translate character") translates or deletes selected characters. It takes two sets of characters as arguments, and replaces occurrences of the characters in the first set with the corresponding elements from the second set. The following one-liner replaces line breaks with spaces:

```
tr "\n" " " < yourfile
```

In other words, many lines are transformed into a single line. The quotation marks are only needed because the newline character and the space are special symbols. tr deals with characters, not strings. The command `tr ab AB` would *not* merely replace any occurrence of the string "ab" with "AB", but it will replace every individual "a" with "A" and every "b" with "B". One finds out about the behavior either by reading the documentation, or by trial and error.

sed is a *stream editor* used for noninteractive editing. The following replaces the word "NaN" with "-9999" everywhere:

```
sed 's/NaN/-9999/g' yourfile
```

The syntax of the s command (as in "substitute") is: s/string/replacement/flag. Without the flag g at the end (which stands for "global"), only the first occurrence in each row would be replaced.

The following prints lines 3 through 7 of a file:

```
sed -n '3,7p' yourfile
```

sed processes a file (or input stream) line by line and makes only one pass. Typical uses of sed are text substitution and selective printing of text files.

Awk is a programming language designed for text processing and often used as a data extraction tool. It may best be described as "a pattern scanning and processing language". Here are a few handy one-liners:

- Print all lines where the entry in the first column is larger than zero:
  ```
  awk '{if ($1>0) print}' yourfile
  ```

- Print the first column and the sum of the third and fourth column of a comma separated table, and replace the commas with spaces:
  ```
  awk -F, '{print $1,$3+$4}' yourfile.csv
  ```

  The option -F specifies that the field separator is a comma, while the default would be a white space.

- Calculate the sum of all entries in a column:
  ```
  awk '{ sum+=$1} END {print sum}' table.txt
  ```

  Awk syntax borrows heavily from C: sum+=... stands for sum=sum+..., curly braces enclose program blocks, and a semicolon ";" separates two commands. In awk, variables are automatically initialized to zero (for numbers) or an empty string. The content in the first pair of curly braces is executed for every line, as it was in the previous examples, and what follows the END statement is executed after the end of the file has been reached. END is *not* the end of a program block; } is the end of the program block. Calculating the average is accomplished with

```
awk '{ sum+=$1; n++} END { if (n>0) print sum / n }'
```

Awk also has a few special variables, such as 'NR' for the number or records (lines). A slightly shorter version for the calculation of the arithmetic mean is

```
awk '{ sum+=$1 } END { if (NR>0) print sum / NR }'
```

Awk understands trigonometric and other mathematical functions. It is essentially a full-fledged language. Awk even accepts multiple input streams.

Perl is a bigger language than **awk** and **sed** and can, in principle, replace both. Here we limit its description to only one example:

```
perl -e 'print reverse <>' yourfile
```

flips the order of rows. The Perl language borrows features from other programming languages, including C, awk, and sed. It is great for string parsing and text editing.

All the examples above were one-liners. It is breathtaking how much can be accomplished by these utilities with a single line, and the possibilities are augmented with longer code. As already pointed out, awk and Perl are programming languages by themselves and many-line code can be written to carry out complex tasks. The code will be interpreted or compiled, as that of other very high-level languages. Appendix A contains related information, including how several commands can be combined within a single line.

Pattern matching: Regular Expression and Glob

A *Regular Expression*, regex, or rexexp is a special text string that describes a search pattern. [09] matches 0 or 9, [0-9] any number from 0 to 9, and [0-9a-fA-F] matches a single digit of a hexadecimal number. Another way to describe regex is as a sequence of characters that design a search pattern. A period or dot . matches any one character. Unix utilities and other tools often recognize Regular Expressions (e.g., grep, sed, awk, and Perl do). For example, grep can search for string patterns with Regular Expressions: grep "[89]" yourfile.

A caret ^ confines the search to the beginning of each line

```
grep "^#" myfile
```

Finds all files that begin with #, and grep -v "^#" strips out all files that begin with this character. A backslash \ turns off the special meaning of a character. The plus sign, +, indicates one or more occurrences of the preceding element. For example, ab+c matches 'abc' and 'abbbc', but not 'ac'.

Another, simpler, form of pattern matching is wildcards, also known as *Glob*s, used in Unix-like environments. A wildcard * matches any string and a question mark ? matches any one character. For example,

```
grep main *.c
```

searches for the string "main" in C programs, or rather in files with extension .c. Appendix A tells more about glob-style patterns.

There is a difference in the meaning of the much-used symbols * and ? in Regular Expression versus globs. In glob, the asterisk matches any number of any characters including none. In Regular Expression, the asterisk indicates zero or more occurrences of the preceding element. Regex `ab*c` matches "ac", "abc", "abbbc", but not "abdc". Glob's `ab*c` does not match "ac", but matches "abc", "abbbc", and "abdc". Something analogous holds for the question mark. In regex, `vapou?r` matches both "vapor" and "vapour". In glob, it would match neither.

These examples demonstrate that text processing utilities are versatile and powerful tools to format, extract, clean up, and even analyze data. Again, "text" refers to data in plain-text format, not prose, although the tools can surely be applied to human language text as well. Some of them also work with binary file formats.

13.3   NETWORK AND STORAGE TECHNOLOGIES

There are three physical types of communication media:

- (copper) wires

- fiber optics, and

- wireless

Wireless (radio) is nowadays implemented with low-cost semiconductor technology, and commonly provides up to 100 Mbps (Megabits per second). Ethernet over copper wires can transfer up to 10 Gbps and can be up to several hundred meters long. Optical fibers have higher transfer rates than copper wires (currently up to about 100 Gbps on a single fiber) and can be used over longer distances. They are more expensive than copper wires, not because of the cost of the cable, but because opto-electronic converters are needed at the beginning and end of each line.

The reader can experiment with the typical download and upload speeds she has available with a bandwidth test, e.g., `http://speedof.me/`. Download speeds are faster than upload speeds. Admittedly, we typically download for more than we upload. Assuming an average download speed of 0.1 Gbps, transferring 1 TB takes 22 hours. Data of the size of 1 PB, a thousand times more than 1 TB, are, hence, for all practical purposes, physically stuck in place. Ironically, the fastest way to move very large volumes of data is to physically transport the disks. This gives rise to the concept that analysis of large data should take place where the data are, not where the user is.

Data transfer can only be as fast as the slowest component, and this bottleneck is usually at the beginning or the end.

The probability of error during data transfer is negligible, because error correction is already applied (with checksum) and if that test fails, the transfer will be repeated.

As already described in chapter 8 there are two common types of non-volatile and re-writable storage media (hard drives): magnetic disks and solid state drives. The former are cheap and large, and the latter more expensive and a bit faster. The absolute speed of transfer from hard drives was already included in Table 8.3.

RAID (Redundant Array of Independent Disks) is a storage system that distributes data across multiple physical drives to achieve data redundancy, fast data access, or a bit of both. For data redundancy, it can be arranged that a copy of one disk's data be distributed on the other disks, such that if any single disks fails, all its data can be recovered from the other disks. For fast data access, sequential bytes can be stored on different drives, which multiplies the available bandwidth. A RAID can start with just a few disks, and some scientists have one under their office desk.

Distributed storage. When data sets are very large, they have to be distributed over many physical storage units. And since every piece of hardware has a failure probability, failure becomes commonplace in a large cloud or data center. Modern file systems can automatically handle this situation by maintaining a replica of all data, and without interruption to normal operations ("hot swappable").

Cloud storage is a form of distributed storage. Data stored in the cloud can be accessed through the internet from anywhere. But clouds are also a level of abstraction or virtualization. We do not have to deal with individual disks or size limits. Another benefit is that cloud storage functions as off-site (remote) backup of data.

The practical situation for the cost of the components is

$$\text{Bandwidth} > \text{Disk} > \text{CPU}$$

These inequalities express that network bandwidth is expensive, so the CPU should be co-located with the disk. In other words, bring the algorithms to the data. Disk space is cheap, so bring everything online. CPU processing power is even cheaper, so no need to pre-process needlessly.

The appropriate outcome given these technology-based relations is large remote compute and storage centers. Today, there are data centers of the size of warehouses specifically designed to hold storage media and CPUs, along with cooling infrastructure. Currently, the largest data centers hold as much as an Exabyte of data.

13.4   WEB SCRAPING AND DATA ARCHIVING

*"You have no idea the number of excuses people come up with to
hang onto their data and not give it to you."*

Tim Berners-Lee

An enormous amount of data is connected through the internet. Web scraping,
web harvesting, or web data extraction refers to copying data from a website
or other internet location, typically onto a local storage medium and often
followed by parsing of these data.

`wget` and `curl` are command line tools that download content from web
addresses without user interaction (Exercise 13.1a). This functionality is also
available with the Python module Scrapy. wget is a powerful scraper for harvesting information from the web. (The "w" in wget stands for "web", and
while several pronunciations are in use, "double u get" is certainly a valid
one.)

For example, ftp://aftp.cmdl.noaa.gov/data/meteorology/in-situ/mlo/ is
a data archive for the Mauna Loa Meteorological Observatory (the famous
observatory where the annual rise of atmospheric CO_2 concentration was first
discovered). Entering this URL into a Web browser will show the content of
the directory, which the hosts have decided to make publicly viewable (or
"listable"). This URL uses the ftp protocol, but https or another protocol
would work equally well. Files in this directory can be downloaded manually,
or the command `wget ftp://aftp.cmdl.noaa.gov/data/meteorology/`
`in-situ/mlo/met_mlo_insitu_1_obop_hour_2017.txt` would download this
one file. The same directory also contains a README file that describes
the meaning (and units) of individual columns, and other metadata, such as
references to related published material. Data and metadata are all in plain
text format, a convenient and appropriate choice for data of this volume.

The same tools allow entire directories to be downloaded, and wget accepts
wildcards for the filenames, but wildcards will only work if the directory is
listable. When a URL directory is not listable, the filenames need to be known
ahead of time (and guessing such a filename would be as difficult as guessing
a password). When a login is required to access data, wget and curl still work,
but need to be configured for noninteractive authentication.

Now that it was described how to download data produced by other people, how can we make our data available to others? Archiving of data enhances
their scientific value. In particular, it allows for combining data from different
sources. The lower the barrier to finding, obtaining, and understanding data,
the better. Data are more helpful when they are *curated* (documented, val-
idated, rigorously formatted, ...), and that requires extra effort by the data
creator. Data by themselves, without any scientific conclusion derived from
them, can be published and cited, giving the data creators the credit they
deserve.

The following features serve the function of an archive: Long-term availability, open accessibility (without pre-approval), standardized formats and file types, peer review, documentation, and citability.

Often research projects are finished only years after the data have been collected, and waiting to archive the data until then is ultimately not the most efficient approach to archiving. Many experienced data archivers find that it is practical to archive promptly, as expressed in the phrase "archive early and archive often".

Longevity of file formats: As technology changes, both hardware and software may become obsolete. And if the past is any guide, they really do. (An infamous example is the loss of tapes from the Apollo 11 Moon landing. They were forgotten and overwritten.) File formats more likely to be accessible in the future are non-proprietary, open, documented standards commonly used by the research community. Where file size permits, uncompressed plain text (ASCII or Unicode) is an excellent choice. Tables can be stored in plain text format (with columns separated by commas, spaces, or tabs), rather than in spreadsheet software specific file formats. Images, sound files, and videos should not require commercial or overly specialized software to be opened.

Exemplary data storage comes from the realm of bioinformatics. The GenBank sequence database is an open access collection of all publicly available nucleotide sequences. The consequence is that anyone who obtains a new sequence can instantly find the genetically closest known relatives.

Recommended Reading: A complete documentation of sed can be found at `www.gnu.org/software/sed/manual/html_node/index.html`. Gnu Core Utilities are described in `www.gnu.org/software/coreutils/manual/coreutils.html`. McKinney, *Python for Data Analysis: Data Wrangling with Pandas, NumPy, and IPython*, provides an introduction into the mechanics of these highly relevant tools. Pandas is a Python module dedicated to data manipulation and analysis.

EXERCISES

13.1 a. Use wget, curl, Scrapy, or a similar tool to download monthly meteorological data from 1991 to 2017 from `ftp://aftp.cmdl.noaa.gov/data/meteorology/in-situ/mlo/`.

   b. From the data obtained, extract the temperatures for 10-m mast height and plot them. (Hint: awk is a tool that can accomplish this.)

13.2 Use awk, Perl, sed, or another utility to write a one-line command that replaces

   a. a sequence of four stars **** with -999.
   b. a sequence of stars of any length with -999.

13.3 What do the following commands do?

a.   `sed 's/^[ \t]*//'`
b.   `awk '{s=0; for (i=1; i<=NF; i++) s=s+$i; print s}'`

In Regular Expression \t stands for a tab or whitespace, and in awk the variable 'NF' is the number of fields (entries in a row).

## CHAPTER 14

Building Programs for Computation and Data Analysis

This chapter is dedicated to practical aspects of program development, relevant especially for large and complex scientific coding projects. In the process, several widely used programming tools are introduced. Appendix B provides a list of code repositories.

14.1 PROGRAMMING

> *"In computer programming, the technique of choice is not necessarily the most efficient, or elegant, or fastest executing one. Instead, it may be the one that is quick to implement, general, and easy to check."*

Numerical Recipes

> *"The single most important rule of testing is to do it"*

Kernighan & Pike

To a large extent the same advice applies for scientific programming as for programming in general: Programs should be clear, robust, and general. Only when performance is critical, and only in the parts where it is critical, should one compromise these principles. Most programmers find that it is more efficient to write a part, test it, and then code the next part, rather than to write the whole program first and then start testing. In software engineering this practice is known as "unit testing."

In other aspects, writing programs for scientific research is different from software development. Research programs often keep changing indefinitely and are usually used only by the programmer herself or a small group of users. Most programs written for everyday scientific computing are used for a limited time only ("throw-away codes"). Under these circumstances there is little benefit to extract the last margin of efficiency, create nice user interfaces, write extensive documentation, or implement complex algorithms. Any of that can even be counterproductive, as it consumes time and reduces flexibility. Better is the enemy of good. That said, the most likely beneficiary of program documentation is our future self.

It is easy to miss a mistake or an inconsistency within many lines of code. In principle, already one wrong symbol in the program can invalidate the result. And in practice, it sometimes does. Program validation is a practical necessity. Some go so far as to add "blunders" to the list of common types of error in numerical calculations: roundoff, approximation errors, statistical errors, and blunders. Absence of obvious contradictions is not a sufficient standard of checking. Catching a mistake later or not at all may lead to a huge waste of effort, and this risk can be reduced by spending time on program validation early on. We will want to compare with analytically known solutions, including trivial solutions. Finding good test cases can be a considerable effort on its own. It is not uncommon, nor unreasonable, to spend more time validating a program than writing the program.

Placing simple print/output commands in the source code to trace execution is a straightforward and widely used debugging strategy (also called "logging"). Alternatively, or in a addition, a *debugger* helps programmers in this process. With a debugger, a program can be run step-by-step or stopped at a specific point to examine the value of variables. When a program crashes, a debugger shows the position of the error in the source code.

Complex programming and modeling tasks involve a large number of steps, and even if making an error at any one step is small, the risk is multiplied through the many steps. That is the rationalization for why it is practically necessary to program defensively and validate routinely.

Programs undergo evolution. As time passes improvements are made on a program, bugs fixed, and the program matures. As the Roman philosopher Seneca put it: "Time discovers truth." For time-intensive computations, it is thus not advisable to make long runs right away. Moreover, the experience of analyzing the result of a short run might change which and in what form data are output. Nothing is more in vain than to discover at the end of a long calculation that a parameter has not been set properly or a necessary adjustment has not been made. Simulations shorter than one minute allow for interactive improvements, and thus a rapid development cycle. Besides the obvious difference in cumulative computation time (a series of minute-long calculations is much shorter than a series of hour-long calculations), the mind has to sluggishly refocus after a lengthy gap.

For lengthy runs one may wish to know how far the program has proceeded. While a program is executing, its output is not immediately written to disk, because this may slow it down (chapter 8). This has the disadvantage that recent output is unavailable. Prompt output can be enforced by closing the file and then reopening the file before further output occurs.

Data can be either evaluated as they are computed (run-time evaluation) or stored and evaluated afterwards (post-evaluation). Post-evaluation allows changes and flexibility in the evaluation process, without having to repeat the run. Numbers can be calculated much faster than they can be written to any kind of output; it is easy to calculate far more than can be stored. For this reason, output is selective.

When data sets accumulate, it is practical to keep a log of the various simulations and their specifications. The situation is analogous to an experimentalist who maintains a laboratory notebook. Forgetting to write down one parameter might necessitate repetition of the experiment. And not labeling a sample can, under unfortunate circumstances, make it useless. Ironically, some scientists keep notes of their model computations not in an electronic file but in a real paper notebook.

To be compelling, a numerical result and a conclusion derived from numerics, needs to be reliable and robust. This expectation is no different from any other branch of scientific inquiry; a laboratory measurement, an astronomical observation, and a theoretical formula are all expected to hold up to scrutiny, because scientists fool themselves often enough. For numerics, being a younger branch of inquiry, standards are less established. Robustness and convergence tests go a long way. We can change a parameter that should not matter to see if we get the same result, and change an input parameter that should matter to see if the expected change occurs. Fitting a graph to data ought to be robust, and when a few of the points are removed, or even half of the points selected at random, the fit parameters should barely change.

Version control systems are used to archive changes to files. Examples are *CVS* (Concurrent Versions System), *Git, Mercurial*, and *SVN*. Source control (that is, version control of source code) is a basic technique of software development. Version control can also serve as backup and, if hosted on a suitable website, is a mechanism for distributing code to users.

When multiple programmers contribute to the same code, additional complexities arise. The standard saying is: "What one programmer can do in one month, two programmers can do in two months." Common practices for team programming include: setting of coding standards, code review (by the whole team or by another member of the team), self-explanatory variable names, and robust/defensive programming that can handle unexpected inputs. For collaborative data analysis, a best practice is to make what we do easily reproducible. Creating a "data pipeline" that defines the workflow may also help. A natural work sequence is: explore the data, analyze the data, automate the pipeline, document the data pipeline.

14.2 SCRIPTING LANGUAGES

Scripts are programs that automate the execution of diverse tasks. Scripts allow several commands that would otherwise be entered one-by-one to be executed automatically, and without having to wait for a user to trigger each stage of the sequence. Scripting languages are most often interpreted rather than compiled.

In the overview of programming languages of Figure 4.1, scripting languages are included in group III. Common scripting languages are Perl, Unix shell scripts (described below), Lua, sed, and awk. We already have encountered the last two in chapter 13. Even Python could be classified as a scripting language, but Python is much more than that.

"Macros" are more or less also scripts, although an application may be able to "record" a macro based on what the user is doing interactively, whereas scripts are written by entering text commands. Either way, the final function is automation of diverse tasks. Closely related are "glue languages", which can be considered a subset of scripting languages.

There are a number of scripting languages and tools; the type of scripts described next is one of the most common and has survived through many decades.

Shell scripts

Shell scripts are specific to Unix-like environments, but not to the Unix operating system per-se. Appendix A provides an introduction to Unix-like environments, which are available on various operating systems. A shell script is a program that runs line-by-line (that is, it is interpreted) in a Unix shell. There are a few flavors of shells: csh (C shell), tcsh (often pronounced "tee-see-shell"), bash, zsh, and others. They all share a common core functionality, although the syntax can vary.

A shell script conventionally starts with `#!/bin/bash` or `#!/bin/tcsh`. The first two bytes '`#!`' have a special meaning and indicate the file is a script and therefore executable, as opposed to a text file not meant for execution. The lines that follow are any commands that are understood by the Unix environment. Common file extensions for shell scripts are `.cmd`, `.sh`, but none is necessary. (The Unix file system has three types of permissions for each file: readable, writable, and executable. A script file needs to be executable, which distinguishes it from a plain text file which is not executable, so it will not be accidentally interpreted as a list of commands.)

An example of a simple shell script is:

```
#!/bin/csh
prog1.out &
prog2.out &
```

which submits two programs as background jobs.

The following shell script moves two files into another directory and renames them to indicate they belong to model run number 4:

```
#!/bin/tcsh
setenv tardir "Output/"  # target directory
setenv ext "dat4"        # file extension
mv zprofile $tardir/zprofile.{$ext}
mv fort.24  $tardir/series.{$ext}
```

Should an error occur during the execution of one of those commands, the script will nevertheless proceed to the next line. For example

```
gcc myprog.c -o prog.out
prog.out &
```

is dangerous. If the compilation fails it will run the executable nevertheless, although it may be an old executable. It would be safer to precede these lines with `rm -f prog.out`, to make sure any old executable with this name is deleted. This is the spirit of defensive programming.

The following merges all files whose filenames begin with 'data' and end with '.csv' using a wildcard:

```
#!/bin/csh
rm -f alldata.csv
foreach i ( data*.csv )
  cat $i >> alldata.csv
end
```

In Unix-like environments >> appends a file, so the file first ought to be deleted to make sure we start with an empty file.

Scripts are also great for data reformatting. The following takes files with comma-separated values (with file extension .csv), strips out all lines that contain the %-symbol with the `grep` command, and then outputs two selected columns from the comma-separated file. Section 13.2 explained `grep` and `awk`.

```
#!/bin/tcsh
grep -v % data814.csv | awk -F, '{print $1,$4}' > h814.dat
grep -v % data815.csv | awk -F, '{print $1,$4}' > h815.dat
```

Shell scripts can have loops, if statements, variables, and arbitrary length. Shell scripts are a great tool to automate tasks, glue various executables together, run batch jobs, and endless other useful tasks.

14.3   DATA-INTENSIVE PROBLEMS

In the modern world of science, data abound. This is largely due to the availability of large and inexpensive digital storage media, as colloquially articulated in Parkinson's law of Data: "Data expands to fill the space available for storage." The saying originally referred to memory, but is now applicable for storage. It is easier and cheaper than ever to store data.

There is so much data that processing it can be a major challenge, a situation often referred to as "Big Data". Big data is a broad and vague term that refers to data so large or complex that traditional data processing methods or technology become inadequate. They may require distributed storage, new numerical methods, or a even a whole ecosystem of hardware and software (such as "Hadoop" which many cloud services use). Various problems can arise when working with large data sets, and next we will discuss a few scenarios.

Data do not fit in main memory: An example of a data format that is conscious of memory consumption is JPEG2000 (or JP2). This image format incorporates smart decoding for work with very large images. It is possible to smoothly pan a lower-resolution version of the image and to zoom into a portion of the image by loading (and decompressing) only a part of the compressed data into memory.

Data take a long time to read: Even if data fit on the local hard drive, reading them could take a while. The data transfer rate from hard drive to memory, and ultimately to the CPU, is limited. (A current technology, typical transfer speeds from a magnetic disk or a solid state drive are 4 Gb/s = 0.5 GB/s.) This is an appropriate point to remember that compressed files not only save storage space, they can also be read and written faster because they are smaller. There are tools that work with compressed files directly. For example, as `grep` searches a file, `zgrep` searches a compressed file, or more specifically a file compressed with gzip.

The text processing utilities (section 13.2) work with small and big data alike, only with large data sets their computational and memory efficiency matters. This efficiency is determined not by the text utility per se, but by its implementation. That said, the more primitive utility `grep` will most likely perform searches faster than the more sophisticated sed, Awk, and Perl. Some text utilities can also take advantage of multiple CPU cores.

sed, for example, can do in-place replacements, with the option `-i`. For example,

```
sed -i 's/old/new/g' gigantic_file.dat
```

does not output a copy of the input file with the text substitutions, as it would without the `-i` option; instead, it overwrites the input file with the output. And to be clear: the in-place option does not do anything for runtime; it merely saves on storage size.

Data do not fit onto the local hard drive: As long as the data can be downloaded, they can be analyzed and then deleted (streamed through). It becomes increasingly important that the data are formatted rigorously and the programs that process them be fault tolerant; otherwise, the data pipeline will break too frequently.

Data are too big to be downloaded: Storage is easier than transfer, hence analysis has to be where the data are not where the user is. This is a game changer, as now the responsibility for the data analysis infrastructure lies with the data host. Cloud computing primarily deals with this issue, as was already described in chapter 13.

To continue with the example of JPEG2000, the standard also provides the internet streaming protocol JPIP (JPEG2000 Interactive Protocol) for efficient transmission of JP2 files over the network. With JPIP it is possible to download only the requested part of an image, saving bandwidth and time. With JPIP it is possible to view large images in real time over the network, and without ever downloading the entire image.

Data may not fit anywhere, or the data are produced in bursts and cannot be stored fast enough (e.g., particle collider experiments). In other words, the data move too fast. This situation has to be dealt with with on-the-fly data analysis, also known as "stream processing." Arithmetic is much faster than writing to disk, so valuable analysis can be done on the fly. Hardware accelerators, such as GPUs (section 9.4), are natural candidates for such a situation.

Big data is not necessarily large data. Data that are complex but insufficiently structured may require extensive effort to be properly analyzed. Here too, the cloud mantra to give users access to the raw data by placing the CPUs close to the disks, enables users to directly work with the data instead of having to wait for sufficient curation or reduced data products by another party.

EXERCISES

14.1 a. Create a large file containing numbers >100 MiB.
     b. Write a program that reads this file and measures the execution time.
     c. Compress the file with one of the many available compression tools, such as zip or gzip. Determine the compression factor (ratio of file sizes).
     d. Write a program that can read the compressed file directly.
     e. Measure the reduction in read time and compare it with the reduction in file size.

14.2 Write a script that validates and merges a set of files. Suppose we have files with names out.0 to out.99, i.e., 100 of them, and each has entries of the form

```
0 273.15
1 260.
```

Write a script that checks that the first column contains the same values in all files and that the entries of the second column are nonnegative. Then merge the files in the order of the numerical value of their file extension (0,...,99), not their alphanumerical value

(0,1,10,11,...,19,2,20,...). This can be accomplished with a Unix shell script, Python, or another scripting language. Validate that the script works, using a small number of input files.

## CHAPTER 15

A Crash Course on Partial Differential Equations

Partial differential equations (PDEs) are differential equations in two or more variables, and because they involve several dimensions, solving them numerically is often computationally intensive. Moreover, they come in such a variety that they often require tailoring for individual situations. Even the same equation can have a variety of boundary conditions. Usually very little can be found out about the solutions of a PDE analytically, so they often require numerical methods. Hence, a chapter is dedicated to them.

There are two major distinct types of PDEs. One type describes the evolution over time, or any other variable, starting from an initial configuration. Physical examples are the propagation of sound waves (wave equation) and the spread of heat in a medium (diffusion equation or heat equation). These are "initial value problems." The other group is static solutions constrained by boundary conditions. Examples are the electric field of charges at rest (Poisson equation) and the charge distribution of electrons in an atom (time-independent Schrödinger equation). These are "boundary value problems." The same distinction can already be made for ordinary differential equations. For example, $-f''(x) = f(x)$ with $f(0) = 1$ and $f'(0) = -1$ is an initial value problem, while the same equation with $f(0) = 1$ and $f'(1) = -1$ is a boundary value problem.

15.1 INITIAL VALUE PROBLEMS BY FINITE DIFFERENCES

As an example of an initial value problem, consider the advection equation in one-dimensional space x and time t:

$$\frac{\partial f(x,t)}{\partial t} + v\frac{\partial f(x,t)}{\partial x} = 0.$$

This describes, for example, the transport of a substance with concentration f in a fluid with velocity v.

If v is constant, then the solution is simply $f(x,t) = g(x - vt)$, where g can be any function in one variable. This is immediately apparent if one plugs this expression into the above equation. The form of g is determined by the initial condition $f(x,0)$. In an infinite domain or for periodic boundary conditions, the average of f and the maximum of f never change.

(A brief explanation of the nature of the advection equation, with $v(x,t)$ dependent on time and space is in order. f is constant along a path $x(t)$, if the total derivative of f vanishes,

$$0 = \frac{df(x(t),t)}{dt} = \frac{\partial f}{\partial t} + \frac{dx}{dt}\frac{\partial f}{\partial x}$$

Hence, $dx/dt = v(x,t)$ describes such paths, and the equation describes the "moving around of material", or, rather, of f-values. In contrast,

$$\frac{\partial f}{\partial t} + \frac{\partial(vf)}{\partial x} = 0$$

with v inside the spatial derivative, is the local conservation law. When a quantity is conserved, changes with time are due to material moving in from one side or out the other side. The flux at any point is vf and the amount of material in an interval of length $2h$ is $2hf$, hence $\partial(2fh)/\partial t = vf(x - h) - vf(x + h)$. In the limit $h \to 0$, this leads to $\partial f/\partial t + \partial(vf)/\partial x = 0$. Since we will take v to be a constant, the distinction between these two equations does not matter, but it is helpful to understand this distinction nevertheless.)

A simple numerical scheme would be to replace the time derivative with $[f(x,t + k) - f(x,t)]/k$ and the spatial derivative with $[f(x + h,t) - f(x - h,t)]/2h$, where k is a small time interval and h a short distance. The advection equation then becomes

$$\frac{f(x,t + k) - f(x,t)}{k} + O(k) + v\frac{f(x + h,t) - f(x - h,t)}{2h} + O(h^2) = 0.$$

This discretization is accurate to first order in time and to second order in space. With this choice we arrive at the scheme

$$f(x,t + k) = f(x,t) - kv\frac{f(x + h,t) - f(x - h,t)}{2h}$$

As will soon be shown, this scheme does not work.

Instead of the forward difference for the time discretization we can use the backward difference $[f(x,t) - f(x,t - k)]/k$ or the center difference $[f(x,t + k) - f(x,t - k)]/2k$. Or, $f(x,t)$ in the forward difference can be eliminated by replacing it with $[f(x + h,t) + f(x - h,t)]/2$. There are further possibilities, but let us consider only these four. Table 15.1 lists the resulting difference schemes.

For purely historical reasons some of these schemes have names. The second scheme is called Lax-Wendroff, the third Leapfrog (a look at its stencil in

TABLE 15.1 A few finite-difference schemes for the advection equation. The first column illustrates the space and time coordinates that appear in the finite-difference formulae, where the horizontal is the spatial coordinate and the vertical the time coordinate, up being the future. Subscripts indicate the spatial index and superscripts the time step, $f_j^n = f(jh, nk)$.

stencil	scheme
O O O O	$f_j^{n+1} = f_j^n - v\frac{k}{2h}(f_{j+1}^n - f_{j-1}^n)$
O O O O	$f_j^{n+1} + v\frac{k}{2h}(f_{j+1}^{n+1} - f_{j-1}^{n+1}) = f_j^n$
O O O O	$f_j^{n+1} = f_j^{n-1} - v\frac{k}{h}(f_{j+1}^n - f_{j-1}^n)$
O O O	$f_j^{n+1} = \frac{1}{2}(1 - v\frac{k}{h})f_{j+1}^n + \frac{1}{2}(1 + v\frac{k}{h})f_{j-1}^n$

Table 15.1 explains why), and the last Lax-Friedrichs. But there are so many possible schemes that this nomenclature is not practical.

The first scheme does not work at all, even for constant velocity. Figure 15.1(a) shows the appearance of large, growing oscillations that cannot be correct, since the exact solution is the initial conditions shifted. This is a numerical instability.

The reason for the instability can be understood with a bit of mathematics. Since the advection equation is linear in f, we can consider a single mode $f(x,t) = f(t)\exp(imx)$, where $f(t)$ is the amplitude and m the wave number. The general solution is a superposition (sum) of such modes. For the first scheme in Table 15.1 this leads to

$$f(t + k) = f(t) - vkf(t)\frac{e^{imh} - e^{-imh}}{2h}$$

and further to $f(t + k)/f(t) = 1 - ikv\sin(mh)/h$. Hence, the amplification factor

$$|A|^2 = \left|\frac{f(t+k)}{f(t)}\right|^2 = 1 + \left(\frac{kv}{h}\right)^2 \sin^2(mh)$$

which is larger than 1. Modes grow with time, no matter how fine the resolution. Modes with shorter wavelength (larger m) grow faster, therefore the instability.

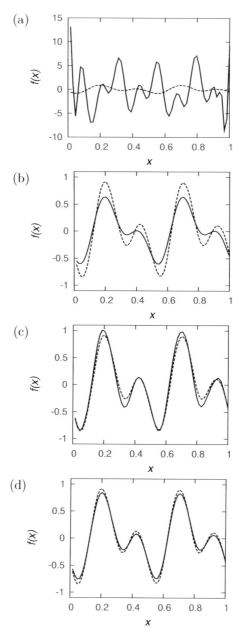

FIGURE 15.1 Numerical solution of the advection equation (solid line) compared to the exact solution (dashed line) for four different numerical methods. All four schemes are integrated over the same time period and from the same initial condition.

The same analysis applied, for instance, to the last of the four schemes yields

$$|A|^2 = \cos^2(mh) + \left(\frac{vk}{h}\right)^2 \sin^2(mh).$$

As long as $|vk/h| \leq 1$, the amplification factor $|A| \leq 1$, even for the worst m. Hence, the time step k must be chosen such that $k \leq h/|v|$. This is a requirement for numerical stability.

The second scheme in Table 15.1 contains f^{n+1}, the solution at a future time, simultaneously at several grid points and hence leads only to an implicit equation for f^{n+1}. It is therefore called an "implicit" scheme. For all other discretizations shown in the table, f^{n+1} is given explicitly in terms of f^n. The system of linear equations can be represented by a matrix that multiplies the vector f^{n+1} and yields f^n at the right-hand side:

$$\begin{pmatrix} 1 & * & & & * \\ * & 1 & * & & \\ & \ddots & \ddots & \ddots & \\ & & * & 1 & * \\ * & & & * & 1 \end{pmatrix}^{n+1} \begin{pmatrix} f_1 \\ f_2 \\ \vdots \\ f_{N-1} \\ f_N \end{pmatrix}^{n+1} = \begin{pmatrix} f_1 \\ f_2 \\ \vdots \\ f_{N-1} \\ f_N \end{pmatrix}^{n}$$

Stars stand for plus or minus $vk/2h$ and all blank entries are zeros. The elements in the upper right and lower left corner arise from periodic boundary conditions. (If we were to solve the advection equation in more than one spatial dimension, this matrix would become more complicated.) The implicit scheme leads to a tridiagonal system of equations that needs to be solved at every time step, if the velocity depends on time. With or without corner elements, the system can be solved in $O(N)$ steps and requires only $O(N)$ storage (chapter 10). Hence, the computational cost is not at all prohibitive, but comparable to that of the explicit schemes. The scheme is stable for any step size. It becomes less and less accurate as the step size increases, but is never unstable.

The third, center-difference scheme is explicit and second-order accurate in time, but requires three instead of two storage levels, because it simultaneously involves f^{n+1}, f^n, and f^{n-1}. It is just like taking half a time step and evaluating the spatial derivative there, then using this information to take the whole step. Starting the scheme requires a single-differenced step initially.

The stability of the last of the four schemes was already discussed. The scheme is first-order accurate in time and second-order accurate in space, so that the error is $O(k) + O(h^2)$. When the time step is chosen as $k = O(h)$, as appropriate for the stability condition $k \leq h/|v|$, the time integration is effectively the less accurate discretization. Higher orders of accuracy in both space and time can be achieved with larger stencils. Table 15.2 summarizes the stability properties of the four schemes.

TABLE 15.2 The same four finite-difference schemes as in Table 15.1, with names and stability properties.

stencil	scheme	stability
○ ○ ○ ○		unconditionally unstable
○ ○ ○ ○	Lax-Wendroff	unconditionally stable
○ ○ ○ ○	Leapfrog	conditionally stable, $k/h < 1/|v|$
○ ○ ○	Lax-Friedrichs	conditionally stable, $k/h < 1/|v|$

Of course, we would like to solve the advection equation with a varying, rather than a constant, velocity. Over small time and space steps the velocity can be linearized, so that the conclusions we have drawn remain practically valid. If the equation is supposed to express a conservation law, the velocity should be inside the spatial derivative and should be discretized correspondingly.

This lesson demonstrates that choosing finite differences is somewhat of an art. Not only is one discretization a little better than another, but some work and others do not. The properties of such schemes are often not obvious, such as numerical stability; it takes some analysis or insight to see them.

15.2 NUMERICAL STABILITY REVISITED

Calculating the amplification of individual modes can reveal the stability of a scheme, but there are also other methods. Series expansion of the last scheme in Table 15.1 leads to

$$f(x,t) + k\frac{\partial f}{\partial t} = f(x,t) - vk\frac{\partial f}{\partial x} + \frac{h^2}{2}\frac{\partial^2 f}{\partial x^2}$$

and further to

$$\frac{\partial f}{\partial t} + v\frac{\partial f}{\partial x} = \frac{h^2}{2k}\frac{\partial^2 f}{\partial x^2}.$$

This is closer to what we are actually solving than the advection equation itself. The right-hand side, not present in the original equation, is a dissipation term that arises from the discretization. The dissipation constant is $h^2/2k$, so the "constant" depends on the resolution. This is called "numerical dissipation." The scheme damps modes, compared to the exact solution. Indeed, in Figure 15.1(d) a decay of the numeric solution relative to the exact solution is discernible. The higher the spatial resolution (the smaller h), the smaller is the dissipation constant, because a suitable k is proportional to h to the first power, and the less is the damping.

There is the intuitive notion that for a PDE-solver to work, the time step must be small enough to include the region the solution depends on—for any numerical integrator of PDEs. For the advection equation, the solution shifts proportionally with time and therefore this causality criterion is $|v| < h/k$ for explicit schemes. This correctly reproduces the stability criterion of the last two schemes in Table 15.1. (In fact, the solution depends only on the "upwind" direction, and it is possible to construct a stable finite-difference scheme that only uses $f(x, t + k)$ and $f(x, t)$ when the velocity is negative, and $f(x, t - k)$ and $f(x, t)$ when the velocity is positive.) In the second scheme of Table 15.1, the implicit scheme, every point in the future depends on every point in the past, so that the causality criterion is satisfied for any time step, corresponding to unconditional stability. The criterion is not sufficient, as the first scheme shows, which is unstable for any step size. In summary, the causality criterion works, as a necessary condition, for all the schemes we have considered.

The causality criterion does not always need to be fully satisfied for numerically stable schemes. Explicit finite-difference schemes for the diffusion equation, $\partial f / \partial t = D \partial^2 f / \partial x^2$, are such an example. It can be shown that in an infinite domain, the solution at time $t + k$ is given by

$$f(x, t + k) = \frac{1}{\sqrt{4\pi Dk}} \int_{-\infty}^{\infty} e^{-(x-x')^2/(4Dk)} f(x', t) dx'.$$

Hence, the solution depends on the *entire* domain even after an arbitrarily short time. The information travels with infinite speed. A simple explicit forward-difference scheme turns out to have the stability requirement $k < h^2/(2D)$ (Exercise 15.1a). Therefore, an integration scheme for the diffusion equation can be numerically stable even when it uses information from part of the domain only. However, the integral from $x' = x - h$ to $x' = x + h$ does include *most* of the dependence as long as $h^2 \geq O(kD)$, so that most of the causal dependence is satisfied for numerically stable schemes.

15.3 BOUNDARY VALUE PROBLEMS BY FINITE DIFFERENCES

Finite differences can also be used to solve PDEs that are boundary value problems, and here we take a brief glimpse at how such problems can be approached. A simple example is the Laplace equation $\nabla^2 f = 0$ in two dimensions $f = f(x, y)$, supplemented by boundary conditions. With the derivatives spelled out individually, the two-dimensional Laplace equation is

$$\left(\frac{\partial^2}{\partial x^2} + \frac{\partial^2}{\partial y^2} \right) f = 0$$

The derivatives can be approximated by finite differences. Since the second derivative is approximately $f''(x) = [f(x-h)-2f(x)+f(x+h)]/h^2$ (chapter 6), a discretization of the Laplace equation around coordinate (0,0) is

$$f(-h,0) + f(h,0) + f(0,-h) + f(0,h) - 4f(0,0) = 0$$

This amounts to a linear system of coupled equations for f, where f may be represented as a one-dimensional vector by renaming the indices; each grid point corresponds to one entry in this vector. In the linear system $Af = b$, the right-hand side b is not zero everywhere because of the boundary conditions. The matrix A is of the form shown in Figure 15.2. This system of equations can be solved for f.

```
-4  1           1
 1 -4  1           1
    1 -4  1           1
       1 -4  1           1
 1        1 -4  1
    1        1 -4  1
       1        1 -4  1
          1        1 -4
```

FIGURE 15.2 Sparse matrix with coefficients that correspond to the finite-difference approximation of the two-dimensional Laplace equation. All blank entries are zero. A sparse matrix of this form is also called "block-tridiagonal".

The dimension of this matrix equals the number of grid points, so it is huge, but it is sparse. If the grid is of size $N \times N$, the dimension of the matrix is N^2 and it has N^4 elements. It would be prohibitive it terms of memory consumption to store the whole matrix. But the non-zero elements are merely five one-dimensional arrays, each of length no more than N^2. In chapter 10 we learned that a tridiagonal matrix can be solved in $O(M)$ steps, where M is the dimension of the matrix. The same holds true for the block-tridiagonal matrix here. The computational cost and the memory requirements to solve the linear system arising from the discretization of the two-dimensional Laplace equation are hence proportional to the number of grid elements, and no more.

The same finite-difference expression for the Laplace equations suggests another method of solution. The center element $f(0,0)$ is the average of its four neighbors,

$$f(0,0) = \frac{f(-h,0) + f(h,0) + f(0,-h) + f(0,h)}{4}$$

This property can be used for an iterative numerical scheme that requires only a few lines of code. Each grid point is assigned the average of its four neighbors. When repeated again and again, the solution relaxes to the correct one. This might not be the computationally fastest way to solve the Laplace equation, and it is not, but it is exceedingly quick to implement. This is a simple version of a "relaxation method." It iteratively relaxes to the correct solution.

When every derivative in a PDE is replaced by a finite difference, the PDE turns into a large system of equations. This very general approach often leads to a workable numerical scheme.

15.4   OTHER METHODS FOR PDEs

The major types of methods for solving PDEs are

- Finite-difference methods
- Spectral methods (and more generally, integral transform methods)
- Finite-element methods
- Particle methods

In finite-difference methods all derivatives are approximated by finite differences. The four examples in Table 15.1 are all of this type.

For spectral methods at least one of the variables is treated in spectral space, say, Fourier space. In other words, spectral methods use spectral representations of the solution. For example, the advection equation with a time-dependent but space-independent velocity, $\partial f/\partial t + v(t)\partial f/\partial x = 0$, becomes $\partial \hat{f}(\kappa,t)/\partial t = -i\kappa v(t)\hat{f}(\kappa,t)$, where κ is the wave number and $\hat{f}$ the Fourier transform of f with respect to x. In this simple case, each Fourier mode can be integrated as an ordinary differential equation. (Also for this specific example, if the boundary conditions are imposed at specific locations in space, it would be cumbersome to translate them into Fourier space. On the other hand, if the boundary conditions are periodic, they are easily treated in Fourier space. As often for the numerical solution of PDEs, implementation of the boundary condition is a major part of the work.) When a derivative is evaluated in spectral space, the approximation error is extremely small, because the stencil is essentially the whole domain.

Fourier series are not the only spectral representation. Representation in terms of Chebyshev polynomials is another widely used type of spectral method.

For finite-element methods (FEMs) the domain is decomposed into a mesh other than a rectangular grid, commonly triangles (or in higher dimensions, polyhedrons) with varying shapes and mesh density. Such grids can accommodate boundaries of any shape and solutions that vary rapidly in a small part of the domain. Mathematically, FEMs are often viewed from the perspective of error minimization, that is, minimizing the deviations from the exact solution, summed over all grid points.

Particle methods represent the solution in terms of fields generated by localized objects. The idea is exemplified by the gravitational field generated by point masses. The gravitational potential Φ obeys a PDE, in empty space the Laplace equation $\nabla\Phi = 0$; only at the location of the masses $\nabla\Phi \neq 0$. The gravitational potential of a single mass m is simply $\Phi = -Gm/|\mathbf{r}|$ (where G is the gravitational constant), and the total potential is the sum over these individual potentials. To calculate the force exerted by a collection of point masses, it is easier to sum the individual forces (or potentials) instead of solving the Laplace equation in all of space. If the masses are not pointlike, the potential can be obtained by integration over the extent of the mass distribution. In mathematics, this is called an "integral formulation" of a partial differential equation. This approach will be described in chapter 16 for the electric potential.

Recommended Reading: LeVeque, *Numerical Methods for Conservation Laws*, provides an introduction to the topic described in its title, especially the one-dimensional case. Brenner & Scott, *The Mathematical Theory of Finite Element Methods* is a classic textbook on this subject.

EXERCISES

15.1 The heat or diffusion equation

$$\frac{\partial f}{\partial t} = D\frac{\partial^2 f}{\partial x^2}$$

describes the evolution of temperature $f(x,t)$ for a (constant) thermal diffusivity D.

a. Carry out a stability analysis for the finite-difference scheme

$$\frac{f_j^{n+1} - f_j^n}{k} = D\frac{f_{j+1}^n - 2f_j^n + f_{j-1}^n}{h^2}$$

and derive the condition for numerical stability.

b. The backward time difference

$$\frac{f_j^{n+1} - f_j^n}{k} = D\frac{f_{j+1}^{n+1} - 2f_j^{n+1} + f_{j-1}^{n+1}}{h^2}$$

leads to an implicit scheme. Show that this scheme is unconditionally stable.

c. The Crank-Nicolson method for the heat equation uses the average of the spatial derivatives evaluated at time steps n and $n+1$:

$$\frac{f_j^{n+1} - f_j^n}{k} = \frac{1}{2}D\frac{f_{j+1}^n - 2f_j^n + f_{j-1}^n}{h^2} + \frac{1}{2}D\frac{f_{j+1}^{n+1} - 2f_j^{n+1} + f_{j-1}^{n+1}}{h^2}$$

Such a superposition of a fully implicit and an explicit method is called semi-implicit. Show that the Crank-Nicolson method is unconditionally stable.

d. Show that the time discretization error for the Crank-Nicolson method is less than for the fully implicit method. (The analogous conclusion was reached for an ODE solver in Exercise 7.5.)

15.2 Implementation and validation of solver for diffusion equation.

a. Carry out Exercise 15.1a.
b. Implement this explicit scheme using periodic boundary conditions.
c. Derive the equation for the spread of a Gaussian function over time.
d. Verify the validity of the numerical solver with an analytic solution starting with a Gaussian.

15.3 a. Write down a finite-difference approximation for the conservation law

$$\frac{\partial f}{\partial t} = \frac{\partial}{\partial x}\left(D(x)\frac{\partial f}{\partial x}\right)$$

with boundary conditions $f(x = 0, t) = 1$ and $\partial f(x = L, t)/\partial x = 1$.
b. Implement the scheme. Use $L = 2$ and $D(x) = 1 + x^2/5$.
c. Verify numerically that after the solution has settled into its stationary form, the flux $F(x) = D(x)\partial f/\partial x$ is the same at every grid point. A correct discretization is *flux-conservative*.

Submit program source code and program output.

15.4 Implement a spectral solver for the advection equation with periodic boundary conditions, and validate the numerical solution with an analytical solution.

## CHAPTER 16

Reformulated Problems

A Nobel prize was awarded for the invention of an approximation method that can calculate the quantum mechanical ground state of many-electron systems. This breakthrough was based, not on the invention of a new numerical method or algorithm, nor on the implementation of one, but on an approximate reformulation of the governing equation. This last chapter illustrates the power of reformulating equations to make them amiable to numerical solution. It also places us in the realm of partial different equations that are boundary value problems. The chapter, and the main text of the book, ends with an outline of the Density Functional Method, for which the prize was given.

16.1 THREE AND A HALF FORMULATIONS OF ELECTROSTATICS

Those familiar with electrostatics, or one of its mathematical equivalents, are well aware that it can be described with rather different but equivalent equations. Here we look at these formulations from the perspective of numerical evaluation.

The electric field $\mathbf{E}$ in a static situation obeys the equations

$$\begin{aligned} \nabla \cdot \mathbf{E} &= \rho/\epsilon_0 \\ \nabla \times \mathbf{E} &= 0 \qquad \text{(formulation 1)} \end{aligned}$$

Here, ρ is the charge density and ϵ_0 a universal physical constant. These are four coupled partial differential equations (PDEs) for the three components of $\mathbf{E}$. Since the curl of $\mathbf{E}$ vanishes, it is possible to introduce the electric potential Φ, and then obtain the electric field as the derivative of the potential $\mathbf{E} = -\nabla\Phi$, thus

$$\nabla^2 \Phi = -\frac{\rho}{\epsilon_0} \qquad \text{(formulation 2)}$$

The potential is a scalar function, and hence easier to deal with than the vector $\mathbf{E}$. This simple reformulation leads to a single partial differential equation, and hence simplifies the problem tremendously. Instead of solving four coupled PDEs, we solve one PDE.

In empty space, the potential obeys $\nabla^2 \Phi = 0$. When the right-hand side vanishes it is called the "Laplace equation". When there is a source term, as in formulation 2, it is called "Poisson equation".

There is yet another formulation of electrostatics. It is well known that the electric potential of a point charge is given by

$$\Phi = \frac{1}{4\pi\epsilon_0}\frac{q}{r}$$

where q is the electric charge and r is the distance between the point charge and the point where the potential is evaluated. For a collection of point charges the potentials add up,

$$\Phi(\mathbf{r}) = \frac{1}{4\pi\epsilon_0}\sum_j \frac{q_j}{|\mathbf{r}-\mathbf{r}_j|}$$

where the sum is over point charges with charge q_j at position $\mathbf{r}_j$. Generalizing further, for a continuous charge distribution the potential can be expressed by integrating over all sources,

$$\Phi(\mathbf{r}) = \frac{1}{4\pi\epsilon_0}\int \frac{\rho(\mathbf{r}')}{|\mathbf{r}-\mathbf{r}'|}d\mathbf{r}' \qquad \text{(formulation 3)}$$

This was a physically guided derivation; with a bit of vector calculus it could be verified that this integral indeed satisfies the Poisson equation, and is thus equivalent to formulation 2. (Those familiar with vector analysis know that the Laplacian of $1/r$ is zero everywhere, except at the origin.) Each of the three formulations describes the same physics. The integral (formulation 3) is an alternative to solving the partial differential equation (formulation 2).

Such methods can be constructed whenever a PDE can be reformulated as an integral over sources. (Another example is the integral solution to the diffusion equation given in section 15.2.) It would be difficult to imagine a situation where Formulation 1 is less cumbersome to evaluate than formulation 2, but whether formulation 2 or 3 is more efficient numerically depends on the situation. If the potential is required at only a few locations in space and if the charges occupy only a small portion of space, the integral may be computationally the more efficient formulation. To put it in the most dramatic contrast: The electric potential of a single point charge can be evaluated with a simple formula, or one could solve the Laplace equation over all space to get the same result. Alternatively, if the source is spread out, it is easier to solve the Laplace equation for the given source distribution once and for all instead of evaluating the integral anew at each location. So, depending on the situation, formulation 2 or 3 is faster.

In addition to exact reformulations, there may be *approximate* formulations of the problem that drastically reduce its computational complexity. An approximation to formulation 3 above is

$$\Phi(\mathbf{r}) = \frac{1}{4\pi\epsilon_0} \frac{q}{|\mathbf{r} - \mathbf{r}'|} \quad \text{with} \quad q = \int \rho(\mathbf{r}')d\mathbf{r}'$$

where a small spatially restricted patch of electric charge is replaced with a point charge. This last formulation has a trivial dependence on $\mathbf{r}$, whereas in the exact formulation 3, the integral needs to be re-evaluated if $\mathbf{r}$ changes. This is exactly what we have done for the gravitational field in section 12.4. The gravitational force (or potential) from a distant star cluster can be approximated by that of its center of mass.

If the approximation is a good one, this can be a game changer. Due to roundoff or computational limitations, no numerical result is exact, so fundamentally an approximation is not necessarily inferior.

(The potential of a localized source can be written in terms of a multi-pole expansion. The monopole potential decays as $1/r$, the dipole potential as $1/r^2$, the quadrupole potential as $1/r^3$, and so on. Exercise 16.1 will demonstrate that the dipole moment of a group of masses vanishes. Hence the error made by replacing a collection of masses with their center of mass decays *two* factors of r faster than the potential. For a collection of electric charges, this is in general not the case. Mass can only be positive, whereas electric charge can have either sign.)

16.2   SCHRÖDINGER EQUATION*

The spatial behavior of microscopic matter, of atoms and electrons, is described by the Schrödinger equation, which is a partial differential equation for the complex-valued wavefunction $\psi(\mathbf{r})$ in a potential $V(\mathbf{r})$. Both are functions of the three-dimensional position vector $\mathbf{r}$. In its time-independent form the Schrördinger equation is

$$-\frac{1}{2}\nabla^2\psi(\mathbf{r}) + V(\mathbf{r})\psi(\mathbf{r}) = E\psi(\mathbf{r})$$

and the wavefunction must be normalized such that the integral of $|\psi(\mathbf{r})|^2$ over all space yields 1, $\int |\psi(\mathbf{r})|^2 d\mathbf{r} = 1$. This is a boundary value problem, which may have normalizable solutions only for certain values of energy E (which is how energy quantization comes about).

The energy is obtained by multiplying the above expression with the complex conjugate ψ^* and integrating both sides of the equation, because the right-hand side becomes simply E.

$$\int \left[-\frac{1}{2}\psi^*\nabla^2\psi + V|\psi|^2 \right] d\mathbf{r} = \int \psi^* E\psi d\mathbf{r} = E \int |\psi|^2 d\mathbf{r} = E$$

For the ground state, the energy E is a minimum, a fact that is formally known as the Rayleigh-Ritz variational principle. An expression for the ground state energy is

$$E_{\text{gs}} = \min_{\psi} \int \left[-\frac{1}{2} \psi^* \nabla^2 \psi + V |\psi|^2 \right] d\mathbf{r}$$

where the minimum is over all normalized complex functions. The principle implies that we can try out various wavefunctions, even some that do not satisfy the Schrödinger equation, but the one with minimum energy is also a solution to the Schrödinger equation.

Here is one way of solving the Schrödinger equation: The wavefunction is written as a sum of functions $\varphi_n(\mathbf{r})$: $\psi(\mathbf{r}) = \sum_n a_n \varphi_n(\mathbf{r})$. If the φ's are well chosen then the first few terms in the series approximate the wavefunction and more and more terms can make the approximation arbitrarily accurate. Adjusting the coefficients a_n to minimize the energy, analytically or numerically, will provide an approximate ground state.

For two electrons, rather than one, the wavefunction ψ becomes a function of the position of both electrons, but ψ still obeys the Schrödinger equation. There is one additional property that comes out of a deeper physical theory, namely that the wavefunction must obey $\psi(\mathbf{r}_1, \mathbf{r}_2) = -\psi(\mathbf{r}_2, \mathbf{r}_1)$. This is called the "Pauli exclusion principle," because it was discovered by physicist Wolfgang Pauli and it implies that the wavefunction for two electrons at the same location vanishes, $\psi(\mathbf{r}, \mathbf{r}) = 0$. For more than two electrons the wavefunction is antisymmetric with respect to exchange of any pair of electrons.

Suppose we wish to determine the ground state of the helium atom. Since nuclei are much heavier than electrons we neglect their motion, and we neglect all magnetic interactions. The potential due to the electric field of the nucleus is $V(r) = -2e^2/r$, where e is the charge of a proton. In addition, there is electrostatic repulsion between the two electrons. The Schrödinger equation for the helium atom is

$$\left[\underbrace{-\frac{1}{2}\nabla_1^2 - \frac{1}{2}\nabla_2^2}_{T} + \underbrace{\frac{e^2}{|\mathbf{r}_1 - \mathbf{r}_2|}}_{V_{ee}} \underbrace{-2\frac{e^2}{r_1} - 2\frac{e^2}{r_2}}_{V_{\text{ext}}} \right] \psi(\mathbf{r}_1, \mathbf{r}_2) = E\psi(\mathbf{r}_1, \mathbf{r}_2)$$

The symbol ∇_1 donates the gradient with respect to the first argument, here the coordinate vector $\mathbf{r}_1$. The expressions are grouped into terms that give rise to the kinetic energy T, electron-electron interaction V_{ee}, and the external potential due to the attraction of the nucleus—external from the electron's point of view, V_{ext}. The Schrödinger equation with this potential cannot be solved analytically, but the aforementioned method of approximation by minimization is still applicable.

A helium atom has only two electrons. A simple molecule like CO_2 has $6 + 2 \times 8 = 22$ electrons. A large molecule, say a protein, can easily have tens of thousands of electrons. Since ψ becomes a function of many variables, three for each additional electron, an increasing number of parameters is required to describe the solution in that many variables to a given accuracy. The number of necessary parameters for N electrons is $(\text{a few})^{3N}$, where "a few" is the number of parameters desired to describe the wavefunction along a single dimension. The computational cost increases exponentially with the number of electrons. Calculating the ground state of a quantum system with many electrons is computationally unfeasible with the method described.

16.3 OUTLINE OF DENSITY FUNCTIONAL METHOD*

It turns out that energies in the N-electron Schrödinger equation can also be written in terms of the charge density $n(\mathbf{r})$,

$$n(\mathbf{r}) = N \int ... \int \psi^*(\mathbf{r}, \mathbf{r}_2, ..., \mathbf{r}_N) \psi(\mathbf{r}, \mathbf{r}_2, ..., \mathbf{r}_N) d\mathbf{r}_2 ... d\mathbf{r}_N.$$

The integrals are over all but one of the coordinate vectors, and because of the antisymmetries of the wavefunction it does not matter which coordinate vector is left out.

For brevity the integrals over all $\mathbf{r}$'s can be denoted by

$$\langle \psi | (\text{anything}) | \psi \rangle = \int ... \int \psi^*(\mathbf{r}_1, \mathbf{r}_2, ...)(\text{anything})\psi(\mathbf{r}_1, \mathbf{r}_2, ...) d\mathbf{r}_1 d\mathbf{r}_2 ... d\mathbf{r}_N.$$

This notation is independent of the number of electrons. We then have

$$\langle \psi | V_{\text{ext}} | \psi \rangle = \int V(\mathbf{r}) n(\mathbf{r}) d\mathbf{r}$$

for the energy of nuclear attraction.

The expression for the total energy becomes

$$E = \langle \psi | T + V_{ee} + V_{\text{ext}} | \psi \rangle = \langle \psi | T + V_{ee} | \psi \rangle + \int V(\mathbf{r}) n(\mathbf{r}) d\mathbf{r}$$

where ψ is a solution to the time-independent multi-electron Schrödinger equation. The ground state can be obtained by minimization over all normalized wavefunctions with the necessary antisymmetry properties. Such trial wavefunctions do not need to satisfy the Schrödinger equation. Of course, the minimization can be restricted to trial wavefunctions with ground state charge density n_{gs},

$$E_{\text{gs}} = \min_{\psi | n_{\text{gs}}} \langle \psi | T + V_{ee} | \psi \rangle + \int V(\mathbf{r}) n_{\text{gs}}(\mathbf{r}) d\mathbf{r}$$

Note that E_{gs} is expressed in terms of n_{gs}, and does not need to be written in terms of ψ_{gs}. The problem splits into two parts

$$E_{\text{gs}} = \min_{n} \left\{ F[n] + \int V(\mathbf{r}) n(\mathbf{r}) d\mathbf{r} \right\}$$

$$\text{and} \quad F[n] = \min_{\psi | n} \langle \psi | T + V_{ee} | \psi \rangle$$

$F[n]$, is called a "density functional"; a "functional" is a function that takes a function, here the electron density $n(\mathbf{r})$, as an argument.

This is known as the Kohn-Hohenberg formulation, and something remarkable has occurred: the energy E_{gs} is a function purely of the charge density $n(\mathbf{r})$ and does not need to be expressed in terms of $\psi(\mathbf{r}_1, \mathbf{r}_2, ...)$, which would be a function of many more variables. The above equations do not tell us specifically how E depends on n, but at least the reduction is possible in principle.

$F[n]$ is the kinetic energy and self-interaction of the electrons expressed in terms of the charge density. It is independent of the external potential V_{ext}; it describes how the electrons interact with themselves. Once an expression, or approximation, for $F[n]$ is found, it is possible to determine the ground state energy and charge density by minimizing E with respect to n for a specific external potential. Since n is a function in 3 rather than $3N$ variables, the computational cost of this method no longer increases exponentially with the number of electrons.

It is essential that a good approximation to $F[n]$ can be found at a reasonable computational cost. This turns out to be possible. These approximations often involve integrals over terms that depend on n and ∇n, "local" quantities.

The wavefunction ψ is not determined by this method, but E provides the energy; n provides the size and shape of the molecule; and changes of the energy with respect to displacements yield the electrostatic forces, such that physically interesting quantities really *are* described in terms of the charge density alone.

This section described two different approaches for the same problem: one using the many-electron wavefunction, and the other the electron density. With a large number of electrons only the latter method is computationally feasible. The problem was split into an expensive part that describes the electrons by themselves and a computationally cheap part that describes the interaction of the electrons with the external potential. Molecules and solids are built by electronic interactions, and the density functional method is one of the most consequential achievements of computational physics; it was awarded a Nobel prize in chemistry. The key progress was achieved not by improvements in numerical methods or computing power, but through a sophisticated mathematical reformulation of the equations to be solved.

EXERCISES

16.1 The gravitational potential of a point mass and the electric potential of a point charge both have potentials of the form $\Phi \propto 1/|\mathbf{r}|$.

    a. A pure dipol in three-dimensional space with cartesian coordinates (x, y, z) consists of two charges $q(0, 0, d)$ and $-q(0, 0, -d)$, separated by a distance $2d$. Show that the potential of a pure dipole decays as $1/r^2$ for $|\mathbf{r}| \gg d$.

b. Show that the dipole moment of a gravitational field expanded around the center of mass is always zero. A dipole moment is defined by

$$p = \int \mathbf{r}' \rho(\mathbf{r}') \, d\mathbf{r}'$$

where the integral is over all space.

16.2 Consider the following equations for u:

a. $-u''(x) = f(x)$, $u(\pm\infty) = 0$, $u'(\pm\infty) = 0$

b. $-k^2 \hat{u}(k) = \hat{f}(k)$

c. $\int_{-\infty}^{\infty} u'v' dx = \int_{-\infty}^{\infty} fv dx$

d. $\min_u \int_{-\infty}^{\infty} \left(\frac{u'^2}{2} - fu \right) dx$

f also decays toward infinity, $f(\pm\infty) = 0$; v is an arbitrary continuous and differentiable function; $\hat{u}$ is the Fourier transform of u. Show that, for "well-behaved" u and f, these are all equivalent. (Note: Each of these formulations motivates an approach for numerical solution.)

16.3 Numerically calculate the ground state energy of the two-electron helium atom.

a. Write down the Schrödinger equation for two electrons, including the electron-electron charge interaction. Both electrons occupy the same spherically symmetric orbital.

b. The one-electron solution is $\psi(r) = b \exp(-2r/a_0)$, where b is a normalization constant and a_0 is the Bohr radius (for the hydrogen atom $\psi(r) \propto \exp(-r/a_0)$). Express the two-electron wavefunction as $\psi(r_1, r_2) = b \exp(-a(r_1 + r_2))(1 + c|r_1 - r_2|)$, with unknown coefficients a and c.

c. Derive the analytic expression for the energy in terms of these coefficients.

d. Numerically minimize the energy with respect to the coefficients.

The Unix Environment

Unix is an operating system, and the many variants of Linux are essentially dialects of Unix. For many decades, including the current decade, Unix-like environments have been a popular choice for scientific, and many other types, of work. Moreover, large computer clusters and clouds often run on a Unix-like operating system. For readers who are not already familiar with the Unix environment, here is a compact introduction.

All major operating systems can provide Unix-like environments. MacOS is based on Unix, and on a Mac the command line interface to Unix can be reached with Applications > Utilities > Terminal. For Microsoft Windows, emulators of a Unix environment are available (e.g., Cygwin and MKS Toolkit). Installing such an emulator makes Unix, and more importantly all the tools that come with it, available also on Windows.

A large number of widely used, and incredibly practical, tools are available within Unix. It is these tools that are of primary practical interest, irrespective of the operating system. In fact, most people who use Unix tools work on computers with operating systems other than Unix or Linux.

Even to this day, the most flexible and fundamental way of working with Unix is through the traditional command line interface, which accepts text-based commands. The most basic Unix commands are often only two or three letters long. For example,

```
cp file1 file2
```

copies file1 to file2. Directory names end in a slash to distinguish them from file names. For example, the command

```
mv file1 Data/
```

moves file1 into directory Data/.

And it gets fancier. For example,

```
scp myfile.txt username@hostname.hawaii.edu:
```

will copy a file onto a *remote* computer or cloud connected through a network, such as the internet. The colon at the end indicates that the second argument is not a file name but a remote computer (that the user must have an account on). Once there was a `rcp` command for "remote copy", but it was superseded by "secure remote copy" or `scp`.

`ssh` (secure shell) provides the important functionality of logging into a remote Unix-based computer:

```
ssh username@ec2-198-51-100-1.compute-1.amazonaws.com
```

for an instance on the Amazon cloud. As mentioned above, `scp` can be used to copy files to and from remote machines, and with the recursive option `-r`, entire directories can be copied.

Many Unix commands understand "Regular Expressions" (described in chapter 13), and even more use "shell globs". For example, `cat *.txt` displays the content of all files with extension `txt`. The command `cat` is an abbreviation for "concatenate" and can be used to display the content of files. The most common glob rules are `*` for zero-or-more characters and `?` for exactly one character. A way to express a choice of one or more strings is `{...,...}`. For example, `ls {*.csv,*.txt}` lists all files with either of these two file extensions.

Command line options. Command options are preceded by a dash. For example, `head` displays the first ten lines of a file, and `head -n 20` displays the first twenty. `ls *.dat` lists all file names with extension `.dat`, and `ls -t *.dat` does so in time order. (Every file has a time stamp which marks the last time the file was modified.) `ls -tr *.dat`, which is equivalent to `ls -t -r *.dat`, lists the content of the directory in reverse time order. Options that are more than one letter long are often preceded by a double dash. For example, `sort -k 2` is equivalent to `sort --key=2` and sorts according to the second column.

Processes and jobs. An ampersand symbol `&` after a command

```
a.out &
```

executes the command in the background instead of in the foreground. A background job frees up the prompt, so additional commands can be entered. It continues to execute even after the user logs out, until it is finished, killed, or the computer goes down.

`ps` lists current processes, including their process id (an integer). `top` displays information about processes. `kill` terminates a process based on its id, and `pkill` based on its name. That way a process that runs in the background can be terminated.

The commands `more` and `less` also display the content of (plain-text) files. After all, less is more. The simplest form of a print statement is `echo`.

Redirect and pipes. Another basic function is the redirect:

```
echo Hello World > tmp
```

writes the words "Hello World" into the file `tmp`. In other words, the "larger than" symbol redirects the output to a file. If a file with that name does not already exist, it will be created; otherwise its content will be overwritten.

```
a.out > tmp
```

sends the output of the program `a.out` to a file. The double `>>` appends. Furthermore, `<` takes the input to a command from a file rather than the keyboard. Without redirects, the standard output is the display, and the standard input is the keyboard. (To be complete, in addition to an input and an output stream, there is also a standard error stream, which by default is also the screen/display. The "larger than" `>` does not redirect the standard error stream, meaning any error messages still end up on the display.)

The output of one command can be sent directly as input to another command with a so-called pipe `|`. For example,

```
a.out | sort
```

sorts the output of the program.

```
a.out | tee tmp
```

feeds the output of `a.out` to the program `tee`. The program `tee` then displays the output *and* stores it in the file `tmp`. More generally, tee is a utility that redirects the output to multiple files or processes.

Redirects and pipes have a similar function; the difference is that redirection sends the output to a file whereas pipes send the output to another command.

Some symbols need to be "escaped" with a backslash. For example, since a space is significant in Unix, a space in a file name "`my file`" often needs to be written as `my\ file` to indicate this is a space in a single argument rather than two separate arguments. Similarly, a Regular Expression, such as `*` may need to be written as `\*`.

Unix tools consist of hundreds of built-in core utilities. A number of useful Unix utilities, such as `diff`, `sort`, and `paste`, have already been mentioned in chapter 13. The text processing utilities described in chapter 13 can be combined in a single line. For example,

```
grep NaN mydata.dat | awk '{if ($1==0) print}'
```

finds all lines that contain NaN and then outputs those lines that also have 0 in their first column.

Environment variables. An environment variable can affect the way a running process behaves. For example, LD_LIBRARY_PATH is the directory or set of directories where libraries should be searched for during compilation. A running process can query the value of an environment variable set by the user or the system. Similarly, environment variables know the number of cores or threads on a CPU and the id of the core a job is running on.

The amount of freely available Linux software is gargantuan. As scientists we may be interested in compilers, such as `gcc`; Python and its many modules; text processing languages, such as `sed` and `awk`; graphing packages such as Gnuplot; tools that convert between file formats, such as `enscript`, `pdftotext`, and ImageMagick; editors, such as vi and Emacs; and a great number of other software.

Recommended Reading: Unix tools are free, and abundant and equally free documentation is available online. Gnu Core Utilities are documented on many websites, including `www.gnu.org/software/coreutils/manual/coreutils.html`. In terms of books, Robbins *Unix in a Nutshell* is a comprehensive guide, that covers basic Unix commands, Unix shells, Regular Expressions, common editors, awk, and other topics.

Numerical Libraries

The *Guide to Available Mathematical Software* at `http://gams.nist.gov` maintains a cross-index of mathematical software packages from numerous public and proprietary repositories.

General Code Repositories:

- *NETLIB* at `www.netlib.org` offers free sources from a myriad of authors.

- *Numerical Recipes*, `http://numerical.recipes/`, explains and provides a broad and selective collection of reliable subroutines. The source code accompanies the book. Each program is available in several languages: C, C++, Fortran 77, and Fortran 90. Code is proprietary but you will get the source code.

- The *Gnu Scientific Library* (GSL) is a collection of open-source C routines, `www.gnu.org/software/gsl`.

- *IMSL (International Mathematical and Statistical Library)* is a proprietary collection of numerical analysis software libraries, available in multiple programming languages.

- *NAG (Numerical Algorithms Group)* Library is also proprietary and available in multiple languages.

General-purpose numerical computing environments include a great number of mathematical functions. Octave makes use of GSL functions. The functions used by SciPy are also available individually (and often not implemented in Python, but in C or Fortran).

A few noteworthy specialized numerical libraries:

- BLAS (Basic Linear Algebra Subprograms) and LAPACK (Linear Algebra PACKage) are highly optimized libraries for numerical linear algebra. They are available at `www.netlib.org/blas/` and `www.netlib`.

`org/lapack/`, but already incorporated in many other software packages.

- An exceptional implementation is FFTW, the "Fastest Fourier Transform in the West" (where "west" refers to Europe's perspective of the location of MIT—where the package was developed—on the East Coast of North America). It is a portable source code that first detects what hardware architecture it is running on and chooses different computational strategies depending on the specific hardware architecture. This way, it can simultaneously achieve portability and efficiency.

- A specialized, refereed set of routines is available to the public from the *Collected Algorithms of the ACM* at `http://calgo.acm.org`.

- The Stony Brook Algorithm Repository, `http://www3.cs.stonybrook.edu/~algorith/`, provides source routines for a number of standard algorithm and data structure problems. It is accompanied by the book *The Algorithm Design Manual* by Steven Skiena.

- The Computational Geometry Algorithms Library (CGAL), `www.cgal.org`, provides algorithms for geometric problems, such as triangulation or intersection of polygons. It is written in C++.

- PETSc (Portable, Extensible Toolkit for Scientific Computation) is a framework for solving partial differential equations, `www.mcs.anl.gov/petsc/`

Answers to Brainteasers

Chapter 1: Analytical & Numerical Solutions

All of them can be solved analytically.

(i) In general only polynomials up to and including 4th degree can be solved in closed form, but this 5th degree polynomial has symmetric coefficients. Divide by $x^{5/2}$ and substitute $y = x^{1/2} + x^{-1/2}$, which yields a polynomial of lower order.

(ii) Any rational function (the ratio of polynomials) can be integrated. The result of this particular integral is not simple.

(iii) $\sum_{k=1}^{n} k^4 = n(n+1)(2n+1)(3n^2 + 3n - 1)/30$. In fact, the sum of k^q can be expressed in closed form for any positive integer power q.

(iv) An iteration of this type has a solution of the form $y_n = c_1 \lambda_1^n + c_2 \lambda_2^n$. The value of λ can be conveniently determined using the ansatz $y_n = \lambda^n$, which leads to a quadratic equation for λ with two solutions, λ_1 and λ_2.

(v) The exponential of any 2×2 matrix can be obtained analytically. The exponential of a matrix can be calculated from the Taylor series. This particular matrix can be decomposed into the sum of a diagonal matrix $D = ((2,0),(0,2))$ and a remainder $R = ((0,-1),(0,0))$ whose powers vanish. Powers of the matrix are of the simple form $(D+R)^n = D^n + nD^{n-1}R$. The terms of the power expansion form a series that can be summed.

Chapter 6: Approximation Theory

Our convergence test in section 6.2 showed that $\|u_{2N} - u_N\| \to 0$ as the resolution N goes to infinity (roundoff ignored). Does this mean $\lim_{N \to \infty} \|u_N - u\| \to 0$, where u is the exact, correct answer?

In mathematics, a convergence test, without knowing the limit, is accomplished with a *Cauchy sequence*. The defining property of a Cauchy sequence is that its terms eventually all become arbitrarily close to one another. More formally, given $\varepsilon > 0$ there exists an N such that for any pair $m, n > N$, $\|a_m - a_n\| < \varepsilon$. We used $m = 2n$, but only because that is all the resolutions we evaluated. In spirit, we meant to use the preceding N. It is not sufficient for each term to become arbitrarily close to the preceding term ($\|a_{n+1} - a_n\| < \varepsilon$), so our previous test criterion is actually failing in this regard. It should include pairs of n and m that neither have a fixed ratio nor a fixed difference. On the other hand, we would not expect that to matter for our specific example. Given the limitation that we can evaluate u_N for a finite number of indices only, the best we can do is to show that $\|u_M - u_N\| < \varepsilon$ with an "unbiased" relation between M and N and a range of M and N.

Whereas every convergent sequence is a Cauchy sequence, the converse is not always true. The Cauchy sequence criterion only implies convergence when the space is "complete." Completeness means our discrete representations of u must be able to represent the solution u as N increases. In other words, a series with infinitely many terms must be able to match the exact result; obviously a series cannot converge to the correct result if it cannot even represent it. For example, if the function u we seek to approximate goes to infinity, then no trigonometric series will ever converge toward it. As long as u_N can at least in principle match the solution, u_N in a Cauchy sequence converges to a u that no longer depends on the resolution, $u_N \to u$.

Finally, we move to the last part of the question "where u is the exact, correct answer." The answer to this is clearly no. Nothing in that formalism requires u to be the solution to an unnamed equation. It could in principle be the solution to a discretized version of the equation only. Such spurious convergence is possible, if rare. In plain language: just because a numerical solution converges, it does not mean it converges to the correct answer, but if it does not converge, then there is definitely something wrong.

Chapter 10: The Operation Count

Show that $\log(N)$ is never large: Suppose N is the largest representable signed 4-byte integer, $2^{31} \approx 2 \times 10^9$. Even with a base-2 logarithm, $\log_2 N$ is only 31. The resolution of an 8-byte floating-point number is about 10^{-16}, hence there are about 10^{16} distinguishable coordinates. It would make little sense to place more than 10^{16} numbers in the interval 1 to 10, because they become indistinguishable anyway. The logarithm $\log_2 10^{16} \approx 16 \times 10/3 \approx 53$. For these reasons, $\log N$ is never a huge factor, or at least $\log N \ll N$, unless N is so small that we would not be concerned with its size anyway.

Chapter 12: Algorithms, Data Structures, and Complexity

Videos that show how the heapsort algorithm is applied to a suit of cards can be found online, for example, at https://www.youtube.com/watch?v=WYII20au_VY. There are several variants of the heapsort algorithm, so these examples might not exactly match what is described in the text.

Bibliography

[1] M. Abramowitz and I. A. Stegun. *Handbook of Mathematical Functions*. Dover, New York, 1970.

[2] L. Blum, F. Cucker, M. Shub, and S. Smale. *Complexity and Real Computation*. Springer, 1998.

[3] S. Brenner and R. Scott. *The Mathematical Theory of Finite Element Methods*. Springer, third edition, 2008.

[4] R. L. Burden, J. D. Faires, and A. M. Burden. *Numerical Analysis*. Cengage Learning, tenth edition, 2016.

[5] T. H. Cormen, C. E. Leiserson, R. L. Rivest, and C. Stein. *Introduction to Algorithms*. MIT Press, third edition, 2009.

[6] P. J. Davis and I. Polonsky. Numerical integration, differentiation, and integration. In M. Abramowitz and I. A. Stegun, editors, *Handbook of Mathematical Functions*, chapter 25, pages 875–924. Tenth Printing, 1972.

[7] L. Devroye. *Non-Uniform Random Variate Generation*. Springer, 1986.

[8] D. Goldberg. What every computer scientist should know about floating-point arithmetic. *ACM Computing Surveys*, 23(1):5–48, 1991.

[9] G. H. Golub and C. F. Van Loan. *Matrix Computations*. Johns Hopkins University Press, fourth edition, 2012.

[10] B. W. Kernighan and D. M. Ritchie. *The C Programming Language*. Prentice Hall, second edition, 1988.

[11] D. B. Kirk and W. W. Hwu. *Programming Massively Parallel Processors*. Morgan Kaufmann, third edition, 2016.

[12] D. E. Knuth. *The Art of Computer Programming, Volume 2: Seminumerical Algorithms*. Addison-Wesley, third edition, 1997.

[13] R. J. LeVeque. *Numerical Methods for Conservation Laws*. Birkhäuser Verlag, second edition, 2008.

[14] W. McKinney. *Python for Data Analysis: Data Wrangling with Pandas, NumPy, and IPython.* O'Reilly Media, second edition, 2017.

[15] M. Metcalf, J. Reid, and M. Cohen. *Modern Fortran Explained.* Oxford University Press, fourth edition, 2011.

[16] D. A. Patterson and J. L. Hennessy. *Computer Organization and Design: The Hardware/Software Interface.* Morgan Kaufmann, fifth edition, 2013.

[17] W. H. Press, S. A. Teukolsky, W. T. Vetterling, and B. P. Flannery. *Numerical Recipes: The Art of Scientific Computing.* Cambridge University Press, third edition, 2007.

[18] A. Robbins. *UNIX in a Nutshell.* O'Reilly, fourth edition, 2005.

[19] R. Sedgewick and K. Wayne. *Algorithms.* Addision-Wesley, fourth edition, 2011.

[20] S. S. Skiena. *The Algorithm Design Manual.* Springer, second edition, 2011.

[21] J. Stoer and R. Bulirsch. *Introduction to Numerical Analysis.* Springer, third edition, 2002.

[22] L. N. Trefethen. *Approximation Theory and Approximation Practice.* Society for Industrial and Applied Mathematics, 2012.

Index